Taxing Energy Use 2015

OECD AND SELECTED PARTNER ECONOMIES

Please cite this publication as:
OECD (2015), *Taxing Energy Use 2015: OECD and Selected Partner Economies*, OECD Publishing, Paris.
http://dx.doi.org/10.1787/9789264232334-en

ISBN 978-92-64-23232-7 (print)
ISBN 978-92-64-23233-4 (PDF)

The statistical data for Israel are supplied by and under the responsibility of the relevant Israeli authorities. The use of such data by the OECD is without prejudice to the status of the Golan Heights, East Jerusalem and Israeli settlements in the West Bank under the terms of international law.

Photo credits: Cover © Keller – Fotolia.com;
© ARTENS – Fotolia.com;
© EpicStockMedia – Fotolia.com.

Corrigenda to OECD publications may be found on line at: *www.oecd.org/publishing/corrigenda.*

Foreword

This report analyses the taxation of energy use in 41 countries, covering 80% of global energy use. It appears at a juncture when many countries struggle to sustain or reconnect with economic growth and face formidable fiscal consolidation challenges. At the same time, concerns over the very high human costs of air pollution are mounting and the urgency of acting to limit greenhouse gases is now abundantly clear.

Energy use is an important source of greenhouse gas emissions and of air pollution. It also is a critical input into production and consumption in modern economies. If deployed effectively, taxes on energy use are a powerful tool to balance the benefits and costs of energy use. Energy use taxes can also play a useful role in fiscal consolidation. What this report tells us, however, is that with current policies energy taxes fail to live up to their potential.

Taxes on energy use influence the price and use of energy. Ideally, end-user energy prices would reflect their environmental impacts to ensure that resources are used most productively and that the negative side-effects of energy use are contained. Taxes can help to achieve this, while also providing incentives to seek alternative, cleaner technologies.

To employ energy taxes more effectively, it is necessary to understand the signals they provide in respect of energy use. This report systematically analyses the structure and level of energy taxes across 41 countries: the OECD countries and seven selected partner economies (Argentina, Brazil, China, India, Indonesia, Russia and South Africa). Effective tax rates, expressed per unit of carbon and per unit of energy, are situated within the energy market structures and other pricing policies in each country, allowing the price signals they send to be better understood.

Our analysis highlights vastly different levels of energy use and taxation among these 41 countries, but also some common patterns. Transport energy is typically taxed at higher rates than other forms of energy use whereas fuels for heating and process use or electricity generation are more likely to be untaxed or taxed at lower rates. Fuels used for similar purposes are often taxed differently, with low rates applying to some of the fuels most harmful for human health and the environment. Tax rates on coal are particularly low.

The picture is not, however, entirely bleak. The awareness about the need to curb negative side-effects of energy use is rising on governments' political agendas, with many, including the selected partner economies, reconsidering price signals and taxes on harmful forms of energy use and investing in renewable sources of energy. This report can serve as a reference for policy makers and analysts to identify reform options to ensure that energy taxes are best adapted to their economic, social and environmental goals – that is, to develop better tax policies for better lives.

Angel Gurría
OECD Secretary-General

Acknowledgements

This report was prepared by Michelle Harding, Johanna Arlinghaus and Kurt Van Dender of the OECD's Centre for Tax Policy and Administration (CTPA) under the guidance of Kurt Van Dender. The underlying spreadsheet models and the graphics were designed and implemented by Michelle Harding. Chiara Martini of Roma Tre University contributed to the research and drafting of two country chapters.

Other staff in the OECD provided important feedback on the report and contributed to its development: Nils-Axel Braathen, David Bradbury, Stéphane Buydens, Anthony Cox, Brendan Gillespie, Isabelle Joumard, Martine Monza, Hayley Reynolds, Pascal Saint-Amans, Jehan Sauvage, Michael Sharratt, Violet Sochay, Ronald Steenblik, Simon Upton and Chan Yang. Staff in the Tax Policy and Statistics Division of the Centre for Tax Policy and Administration also contributed to the shaping of the report. The authors would like to thank Venkatachalam Anbumozhi, from the Economic Research Institute for ASEAN and East Asia, and Luciano Caratori, from the Instituto Argentino de la Energía "General Mosconi", who provided valuable feedback on the chapters for Indonesia and Argentina, respectively. The authors would also like to thank officials from the governments of the partner economies included in the report for the information and revisions they provided.

The report was discussed at various stages in its development by the OECD's Joint Meetings of Tax and Environment Experts. It was also discussed at, and approved for publication by, the Committee on Fiscal Affairs and the Environment Policy Committee. Invaluable information, comments and other input concerning the report were provided by Delegates to the Joint Meetings and their colleagues in national and sub-national government administrations.

Table of contents

Figures

Executive summary

Taxes can integrate the costs of environmental and health damage and of climate change associated with energy use into its price. There is strong evidence that taxes are one of the most cost-effective ways to curb these negative side-effects of energy use. To what extent does this recognition drive the actual practice of energy taxation? Based on the systematic analysis in this report of effective tax rates on energy use in 41 countries, the answer is readily apparent: with very few exceptions, taxes on energy use do not reflect its negative side-effects. The policy implication is that even where taxes on energy use are also determined by other policy objectives, there is considerable scope for improving their effectiveness.

This report describes in detail the structure and the level of energy use taxes in Argentina, Brazil, People's Republic of China, India, Indonesia, Russian Federation and South Africa, countries that represent a large and growing share of global energy use and carbon emissions. It also draws on *Taxing Energy Use: A Graphical Analysis* (OECD, 2013b) to provide a comparative analysis of energy use and taxation in the 41 countries analysed in both reports, namely all OECD countries and the seven countries just listed. Together, these countries accounted for just over 80% of global energy use and for nearly 84% of carbon emissions from energy use in 2009.

Countries differ widely in terms of their energy mix, the energy intensity of their economies and how energy is taxed. The overall economy-wide level of energy taxation ranges from just over EUR 0 per GJ and per tonne of CO_2 in Indonesia and Russia to EUR 6.58 per GJ in Luxembourg, and EUR 107.3 per tonne of CO_2 in Switzerland. Countries with higher per capita GDP tend to both use more energy per capita and to tax energy use at higher effective rates. Higher effective tax rates on energy also correlate with lower carbon intensity of the economy, although other factors such as energy prices or subsidies and resource availability can affect carbon intensity, and relatively more so where taxes on energy use are lower.

While the level of energy taxation differs strongly across countries, the structure of these taxes reveals some common features. Taxes on energy used in transport are consistently higher than those on other types of energy use. Oil products are taxed markedly more frequently and heavily than other energy sources, with tax rates of just under EUR 14 per GJ in the United Kingdom to less than EUR 5 per GJ in several countries, particularly in Asia and the Americas. Other fossil fuels are taxed at lower rates and are more commonly untaxed. Taxes on natural gas are regularly lower than taxes on oil products, and taxes on coal in particular are often low or zero. The high tax rate on oil products is partially due to the higher tax rates on transport fuels, as transport is heavily reliant on oil products and represents a large share of the total use of oil. Nevertheless, taxes on oil products for non-transport use are also relatively high.

The overall landscape of energy taxation in most countries does not correspond well with the features commonly associated with effective environmental taxation. Taxes on energy use for heating and process energy and taxes on energy used to generate electricity are usually too low to reflect negative side effects and they are frequently lower for more polluting fuels, notably coal. Taxes on road transport are high and may in some cases reflect negative side effects of road transport, but the taxes do not align well with the strong variation of these negative impacts over time and place. For example, pollution and congestion costs are high in cities during peak hours, but fuel taxes are the same as in off-peak hours in rural areas. Regardless of the level applied, the differentiation of taxes on fuels used for similar purposes can be (and frequently is) counterproductive: effective tax rates on diesel are in almost all countries analysed lower than these on gasoline for road use, despite the greater harm associated with diesel use.

In addition to specific energy taxes, many other government policies influence energy prices and usage patterns. These include non-tax policies, differential rates of value-added taxes, and tradable carbon permit systems. These policies can enhance or undermine the use of energy taxation to influence energy prices. For example, among the 41 countries included in the analysis, nineteen apply a differential VAT rate to certain energy products. These are almost uniformly concessionary, so that energy use becomes cheaper relative to other forms of consumption, running counter to effective environmental taxation.

The low tax rates on many harmful forms of energy, and the existence of other measures that provide countering signals, strongly suggest that with few exceptions the 41 countries considered in this report do not harness the full power of energy taxes to reduce environmental harm in a cost-effective way. There is evidence, however, that awareness of the negative side-effects of some sources of energy use, and interest in taxes on energy for managing them, is rising in many countries, who are reconsidering energy taxation and pricing policies and are investing in renewable energy sources Such policies are indispensable to the promotion of sustainable development, which will require accommodating strong growth of energy use while containing its negative side-effects. In this context, reconsidering the structure and level of taxes on energy can assist countries in pursuing their economic, social, and environmental objectives as effectively as possible.

Part I

Overview

Taxing energy use in the 41 countries

This chapter discusses the reasons for energy taxation, patterns of energy use and pricing policies in the 41 countries included in the report. It outlines the methodology used to develop graphical profiles of energy use and taxation in each country and to conduct the cross-country analysis. It then presents the results of the cross-country analysis, considering general trends in energy use, the taxation of energy used in different sectors (transport, heating and process use and electricity generation), economy-wide tax rates on energy, and the links between taxes, GDP and energy use per capita. The final part of the section draws conclusions on the impact of current energy pricing and tax measures.

The statistical data for Israel are supplied by and under the responsibility of the relevant Israeli authorities. The use of such data by the OECD is without prejudice to the status of the Golan Heights, East Jerusalem and Israeli settlements in the West bank under the terms of international law.

1. Introduction

This report analyses and describes the level and the structure of taxes on energy use in 41 countries: the 34 OECD countries and seven selected partner economies; Argentina, Brazil, the People's Republic of China (hereafter "China"), India, Indonesia, the Russian Federation (hereafter "Russia"), and South Africa.

To set the stage for the in-depth analysis in subsequent chapters, this introductory chapter begins by describing the motivation of the work undertaken, in Section 1.1. Section 1.2 briefly discusses why energy use is taxed in practice and explains in some more detail the case for using taxes as environmental policy instruments. Section 1.3 provides a quick overview of patterns of energy use in the 41 countries covered, which together account for 80% of world energy use. It illustrates that the selected partner economies represent a major and quickly growing share of global energy use. Section 1.4 discusses relations between energy taxes and prices, in order to provide insight into the weight taxes on the consumption of energy – which are the focus of the report – have in the formation of the price of energy.

The other sections of Part I are as follows. Section 2 provides a detailed discussion of the way the graphical profiles of the taxation of energy use in Part II of this report (for the selected partner economies), and in Part II of *Taxing Energy Use – A Graphical Analysis* [for OECD countries, (OECD, 2013b)], are structured. The section also explains the underlying methodology and data sources. Section 3 contains an in-depth discussion of the results from a systematic cross-country analysis, examining the tax base and the tax rates across all 41 countries, on an economy-wide basis, and by sector, user type and fuel.

1.1. Motivation

Energy use is crucial for the patterns of production and consumption that characterise modern economies. Ensuring sufficient and secure energy supply and avoiding excessive energy use and limiting negative side effects on health, the environment and the climate is a critical socio-economic process, in which markets and policies are closely intertwined. Price signals are a key feature of this process. Prices influence the amount and type of energy that different users will demand, and they affect how much producers are willing to supply, now and – via investments in capacity – in the future. Prices therefore also affect the health, environmental and climate impacts of energy use. Taxes on the consumption of energy are one tool that governments can use to influence prices, but they are not the only tool: governments can intervene more directly in markets to set or regulate prices using a variety of mechanisms.

Taxes on energy use generate government revenue, a fact well recognised by governments and with considerable impact on the current landscape of energy taxes in many countries. Expenditure on energy use can form a substantial share of total household expenditures, and governments often intervene to contain spending on energy in order to limit inequality or reduce poverty. Energy costs also represent a large share of total costs for some types of firms, and this can inspire public policy choices to limit energy costs in order to support firms' international competitiveness. Taxes are also environmental policy instruments. Since taxes affect the prices of energy products, they can help to steer users' choices on what and how much energy to use. As argued in the next section, taxes on the consumption of energy are among the most cost-effective policy instruments to integrate consideration of the health, environmental and climate costs of energy use into the decision-making of households and firms.

For all these reasons, understanding the structure and level of energy taxes is indispensable for informed policy discussion about energy use. Such policy discussions are a key part of debates on how policy can best support "green growth" as well as contributing to wider policies including industrial, employment, social welfare and health policies.

Taxing Energy Use – A Graphical Analysis was published by the OECD in 2013, with the aim of providing a systematic yet straightforward description of taxes on energy use in OECD countries. The current report describes the structure and the level of taxes on energy use in Argentina, Brazil, China, India, Indonesia, Russia and South Africa. These selected partner economies of the OECD account for 36% of global energy use in 2009, and this share is set to grow quickly along with the weight of these countries in the world economy. Adding the selected partner economies to the set of countries analysed in the 2013 publication results in coverage of 41 countries and just over 80% of world energy use, and 84% of carbon emissions from energy use, in 2009. The present report adopts the same methodological approach as *Taxing Energy Use – A Graphical Analysis* (OECD, 2013b) in that it combines detailed data on energy use (the tax base) with newly-collected information on taxes on energy use, including reported tax expenditures, to produce graphical profiles of the taxation of energy use. The methodology underlying the analysis, and the emphasis in the discussion of the results, are adapted to some specific characteristics of the taxation of energy use found more strongly in the seven countries on which this report focusses than in the OECD countries covered in the 2013 publication. For each of the selected partner economies, graphical profiles of energy use and taxation are constructed that serve three broad goals:

- to understand the composition of energy use and the associated carbon dioxide emissions (CO_2) in each country;
- to illustrate the structure of energy taxation in each country, including:
  - the coverage of the various tax bases related to energy consumption;
  - the effective tax rates in energy and carbon terms that apply to different fuels, uses of fuel, and fuel users;
  - the various tax expenditures that are provided; and
- to help to establish a foundation for analysis of appropriate tax settings on energy.

1.2. Why countries tax, or subsidise, energy use

Together, *Taxing Energy Use – A Graphical Analysis* (OECD, 2013b) and the present report cover 41 countries. All these countries impose broad-based consumption taxes, in the form of VAT or retail sales taxes. A small number of goods and services are frequently subject to specific taxes or subsidies, and the various types of energy are prime examples. Specific taxes on energy are often levied as excise taxes (the focus of this publication) or through differentiation of VAT rates (see Section 1.4). Whereas excise duties increase the price of energy, differential VAT treatment most often means a VAT rate lower than the standard rate and therefore a relatively lower price of energy.

Governments also intervene in other ways, for example through price regulation, to alter energy prices. Price regulation can help to avoid abuse of market power (by aligning prices with short run or long run marginal costs, or average costs), but it can also be used to keep prices below costs (see Section 1.4 for an overview of such mechanisms in the selected partner economies).

There are several reasons why governments intervene to tax or subsidise energy use specifically:

- Taxes can be used to integrate the environmental, health and climate costs of pollution in energy prices, so that energy users will take these costs into account when deciding how much and what energy to use ("internalising external costs"). Tradable permit systems can produce similar results. Box 1 develops the argumentation in some detail.
- Putting specific taxes on energy products often increases the price relative to other goods. This is economically efficient in as far as the demand for energy is relatively inelastic, i.e. demand does not fall strongly when prices increase. The advantage is that market outcomes are not strongly distorted by the tax, so that revenue is raised at a relatively low economic cost. A less elastic tax base is appealing from a revenue-raising point of view but not so much from an environmental point of view, as taxes will reduce pollution less where demand is less elastic. This trade off needs to be considered with the relative weight of revenue-raising and environmental protection objectives in mind.
- In some countries, some types of energy use are subject to a specific tax or charge of which the revenues are hypothecated for particular types of spending. Motor fuel tax revenues, for example, are sometimes reserved for spending in the transport sector. The tax then is akin to a user charge, even if it does not always reflect marginal costs. Where these taxes apply directly to the use of a unit of energy, they are included in our analysis, as their impact is to increase relative prices of energy products, regardless of their stated intent.
- Governments may choose to introduce preferential tax treatment for some types of energy use or for some types of users. They can go further and provide net subsidies for energy use. Such measures are often motivated by distributional or poverty concerns where household use is concerned, and on competitiveness grounds for commercial energy use. Containing inflation is another potential motivation for controlling energy prices, as is stimulating economic development.

In practice, government policies on energy taxation, subsidisation and pricing will be affected by all of the factors mentioned, with the weight of the different factors changing over time and dependent on local constraints and priorities. Economic analysis tends to emphasise the advantages of specific energy taxes as instruments for environmental policy and to some extent as revenue raising instruments (see Box 1). Economic analysis tends to be critical of hypothecation because of risks of misallocating public funds. It also tends to be critical of tax or subsidy policies that keep energy prices for households low, suggesting that distributional goals are better achieved through other means. Competitiveness concerns matter for public policy, but evidence suggests that energy price increases have only limited impacts on firm competitiveness even in energy-intensive sectors (see for example Arlinghaus, 2015).

Box 1. **Why taxes are among the best environmental policy instruments**

Taxes often are levied to raise government revenue, and where this is their principal objective, behavioural responses by taxpayers are usually undesirable. In other cases, including environmental taxation, changing behaviour (to reduce pollution) can be a policy objective, along with revenue raising. Environmentally related taxes are not levied for environmental reasons alone, but – as is explained below – taxes are effective instruments for pursuing environmental objectives.

The environmental, health and climate impacts (in short, pollution) of energy use are not directly borne by producers and consumers, so these costs are not taken into account in decisions based on market prices: these costs are external to the market. The result is that unregulated market outcomes lead to too much pollution, and public policy is needed to improve upon the market outcome by reducing pollution. Governments can intervene with various policy instruments, including taxes, cap-and-trade systems (tradable permits), emission standards, direct technology requirements and restricting the level of pollution-generating activity.

Taxes or auctioned tradable permits tend to outperform other environmental policy instruments in terms of cost-effectiveness. This is because putting a price on pollution provides polluters with incentives to find the cheapest ways of reducing their tax bill. They can reduce the level of the pollution-generating activity or invest in less pollution-intensive ways of carrying out the activity. Alternative instruments, for example energy efficiency standards, imply more prescriptive policy decisions on how to reduce pollution, and given asymmetrical information and heterogeneity among economic agents, the proposed solutions risk not being cost-effective. The economic agents carrying out the pollution-generating activity are better informed than the government about how they can cut pollution, so they are better placed to choose the cheapest option under a regulation-based intervention. Since economic agents differ, the best options can differ as well. For example, some households would be better off by responding to a higher fuel tax by investing in more fuel efficient cars, whereas others would primarily respond by driving less. A fuel economy standard, however, would force the second household to (also) invest in fuel economy, even though this would not be their preferred response. Furthermore, once polluters comply with an energy efficiency standard or a cap on emissions, they do not have an incentive to further reduce pollution, whereas with a tax they have an ongoing incentive to reduce pollution.

Market-based instruments have strong appeal on theoretical grounds and there is evidence that they often work better in practice than other policy instruments (see e.g. OECD, 2013c). Nevertheless, direct regulation, for example with efficiency or emission standards, can be useful in particular circumstances, either in combination with market-based instruments or instead of them. One complication with the use of taxes is that it may be difficult to tax pollution directly and that taxes have to be levied on activities or types of consumption that are more or less strongly related to pollution. When the correlation is weak, taxes become less effective and the relative appeal of direct regulation rises. Fuel taxes, for example, can very accurately reflect the carbon content of fuels and therefore the marginal contribution of fuel use to climate costs, but they correlate less directly with emissions of local pollutants and still less with the ultimate pollution costs resulting from such emissions. Emission standards for local pollutants can usefully complement fuel taxes, but the case for fuel economy standards is weaker. Furthermore, designing effective emission standards is not easy, with e.g. the risk that emission profiles differ substantially between test- and real-world conditions. Using standards to cut pollution is more likely to work well in the early stages of abatement, when pollution is high and cheap technological approaches to reduce it are available. Market-based approaches become more attractive when abatement costs rise and across-the-board measures should make way for more decentralised abatement choices.

Firms often have market power, and this allows them to raise prices above marginal production costs. Producers in the energy sector, where technology often requires large scale operations and where barriers to entry are relatively high, are prone to displaying market power. When market prices exceed marginal production costs, it would seem that the price already covers part or all of the external costs of pollution.

Box 1. **Why taxes are among the best environmental policy instruments** *(cont.)*

The tax needed to internalise the costs of pollution would then be smaller than in a perfectly competitive market. This conclusion, however, is not straightforward. Market power is pervasive throughout the economy, and except in cases where it is particularly strong, prices including mark-ups may not be very far from prices that are as efficient as practically possible. Perfect competition, which would lead to prices equal to marginal costs, is a limiting case that can be approximated but rarely attained in practice. Furthermore, mark-ups can also help pay for fixed costs, and if they do so in ways that are less distortive than alternative ways of funding fixed costs, then mark-ups are efficient in a second-best sense. If market prices are as efficient as possible, then seeking to set taxes or permit prices equal to marginal external costs remains a sensible principle.

In cases where market power is strong, public policy is needed to limit abuse of market power. This can take the form of competition or anti-trust policies and in some sectors (e.g. electricity, utilities) more active regulatory policies or public provision can be justified. In each case, welfare-oriented policy will seek to set prices that align broadly with marginal costs with possibly mark-ups to cover fixed costs and retain investment incentives. As long as such policies are in place and are effective, environmental taxes should still be roughly equal to marginal damage and they should be levied on top of producer prices. Curbing pollution with instruments that generate government revenue (e.g. taxes and auctioned tradable permits) is often preferable to doing so with instruments that do not (e.g. grandfathered tradable permits) or that cost public money (e.g. subsidies for pollution abatement), even if all approaches would yield the same environmental outcome. The reason is that taxes and auctioned permits provide government revenue that can be redirected to more socially or economically advantageous uses, e.g. reducing more distortive taxes, increased expenditure in priority areas, or debt reduction.

Aligning taxes on energy use with its marginal external costs requires estimates of these marginal costs for different forms of energy under different use scenarios. Producing reliable estimates is difficult and costly, and with few exceptions, no "off-the-shelf" information is available for particular situations. A distinction can be made between bottom-up and top-down estimates. Bottom-up estimates model the processes and activities that generate the external cost and combine them with estimates of the economic cost of the impacts. This method provides reliable insight into the strong dependence of some marginal external costs on the time when and place where the externality-generating activity takes place. Air pollution costs, for example, do not only depend on tailpipe or smokestack emissions but also on how these emissions interact with ambient concentrations of other substances, on weather and geographical characteristics, and on how many people are exposed to the resulting level and nature of air pollution. Bottom-up estimates, however, are costly to produce and are not systematically available (see Ricardo-AEA, 2014, for a recent application to the transport sector in Europe).

The difficulty and cost of producing bottom-up estimates of marginal external costs explains why attempts are made to combine bottom-up evidence with indicators of local circumstances to produce top-down estimates of marginal external costs (see e.g. Parry et al., 2014). These estimates may not necessarily be sufficiently precise or may be too aggregated to provide guidance on how large environmental taxes really ought to be, but they do show what directions of environmental tax reform are desirable and in that sense provide very useful context for the interpretation of the evidence on current tax profiles gathered in this publication.

1.3. *Patterns of energy use, cross-sectional and time trends, and country heterogeneity*

The seven selected partner economies represent a large and growing share of global energy use and carbon emissions, as shown in Figure 1. In 2009, the year for which energy use is taken for this publication, they accounted for 36% of world energy use. The OECD countries' share in world energy use was 44.1%. This means that the database underlying

Taxing Energy Use – A Graphical Analysis (OECD, 2013b) and the present publication now covers (just over) 80% of global energy use. The seven selected partner economies include large economies and significant energy consumers. China was the world's largest energy user in 2009, followed closely by the USA. In third and fourth place are two more selected partner economies, India and Russia. The amount of energy used in India and Russia was similar in 2009, at about 30% of the level of energy use in China and the USA. To give further orders of magnitude, Brazil consumed about as much energy as France, and Indonesia consumed slightly less. South Africa is comparable to Italy in the amount of energy used in 2009, and Argentina to Poland or the Netherlands.

Figure 1. **Composition of world energy use**

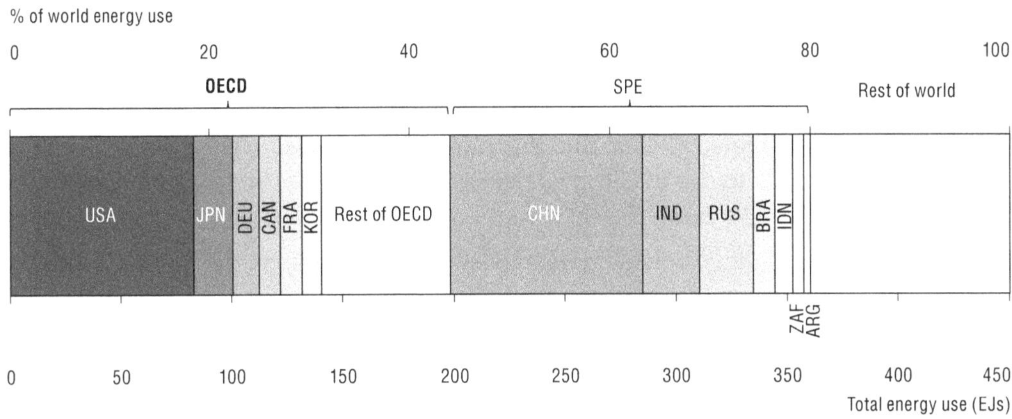

Source: OECD calculations, based on IEA (2014), *IEA World Energy Statistics and Balances* (database), http://dx.doi.org/10.1787/data-00513-en. 2009 information is shown for compatibility with the graphical profiles in Part II of the report.

StatLink http://dx.doi.org/10.1787/888933205545

The shares of energy consumption illustrate the relevance of including the selected partner economies in *Taxing Energy Use*. Furthermore, and as is well known, the weight of the selected partner economies in the world economy is rising and will continue to rise over the coming decades. OECD long-term baseline projections (Johansson et al., 2013) suggest that China's share in global economic output (in purchasing power parity terms) will rise to around 28% in 2030 and 29% in 2050, from 19% in 2013. India's share will rise from 7% in 2013 to 11% in 2030 and 16% in 2050. On the basis of current membership, the OECD's share in global economic output is expected to decline from 62% in 2013 to 49% in 2030 and 43% in 2050. As will be seen, energy use has been strongly dependent on per capita GDP in the past, and if this trend continues, then the share of the selected partner economies in global energy use is set to rise strongly. Part of the increase in energy use is related to transport. The share of transport energy use in total energy use is currently relatively low in the selected partner economies, and it is likely to rise quickly.

The remainder of this section considers the characteristics and the evolution over time of energy use and carbon emissions in Argentina, Brazil, China, India, Indonesia, Russia and South Africa. To provide context, the evolution over time is briefly compared to cross-sectional trends[1] observed in the set of 41 countries formed by the OECD countries and the selected partner economies, and to time trends in four "reference" countries (Netherlands, Spain, United Kingdom and United States, which differ markedly between them in their energy-carbon emission structure[2] but all have very high per capita income). The time series starts in 1990 and ends in 2010, 2011 or 2012, depending on latest available information for the variable considered. The cross-section is for the year 2009. The following stylised facts emerge:

- Rising per capita GDP is associated with higher energy use, expressed in GJ per capita, in the selected partner economies and on average in the cross-section. The association is stronger in the cross-section than in the time series, perhaps because some very high income economies are also very energy-intensive. The association is not positive on average for the four reference countries, as two of them combine GDP growth per capita with declining energy use per capita and the other two display limited or zero growth in energy use while GDP per capita rises. This is suggestive of a transition to less energy-based growth models in economies where per capita income is high.
- Rising per capita GDP is negatively correlated with the energy intensity of GDP (GJ/GDP). This holds in the cross-section and in the time pattern for the selected partner economies (except Brazil) and the four high income economies. Hence, whereas energy use per capita tends to rise as incomes rise (except possibly at very high incomes), the energy intensity of GDP tends to decline.
- Rising per capita GDP correlates positively with CO_2 emissions per capita, except in South Africa, Russia, and the four reference economies where emissions are mostly flat or decline while per capita income increases, over the period considered. The positive correlation in most selected partner economies is the consequence of rising energy use per capita and mostly time-invariant carbon-intensities of energy use.

The stylised facts provide insight into broad trends, but they hide considerable heterogeneity among countries, heterogeneity which is not due to differences in income levels or other indicators of economic development. Chapter 3 analyses patterns of energy use and taxation, and explores their connections with economic characteristics. Some observations include:

- Among the selected partner economies, India and Indonesia have the lowest per capita incomes, and China stands out by its particularly fast income growth, by which it is rapidly moving from the lowest to the median income levels in the selected partner economies.
- Among the selected partner economies, Russia's per capita energy use is particularly high. It is, however, lower than per capita energy use in the USA and comparable to the level of the Netherlands. South Africa's energy use per capita is high too among selected partner economies but it is not very different from the level in Spain or, in recent years, the United Kingdom. Income differences alone clearly do not explain all these differences. Per capita energy use is low in India and Indonesia, even after controlling for low incomes there. Energy use per capita in China starts to grow more quickly as of 2002, around the same time when per capita income growth accelerates.
- As indicated, the energy intensity of GDP (GJ/GDP) declines in all countries except Brazil, from 1990 to 2012. However, it declines from lower levels in the four reference economies, and the rate at which it declines appears not to differ strongly between the reference countries and the selected partner economies (with the exception of China, where the decline is particularly fast, and Brazil, where there is no decline).

1.4. *Taxes and prices of energy use*

This report analyses taxes on energy use in 41 countries: the seven selected partner economies is described in this report and the 34 OECD countries described in *Taxing Energy Use – A Graphical Analysis* (2013b). As discussed in Box 1, taxes on pollutants are highly effective policy instruments. Taxes on energy use can approximate taxes on pollutants more or less directly (e.g. more directly for carbon, less directly for local pollutants), and are among the preferred instruments to include external environmental costs in prices where that is desirable. Among many factors, the prices of energy use depend on other public measures, including regulation, other tax policies and carbon trading mechanisms. This section provides an overview of how taxes, tax expenditures, and other policies shape energy prices. The goal is not to be comprehensive but to place taxes into the broader picture of energy pricing policies.

Section 1.4.1 summarises the main results of this report concerning energy taxation. Section 1.4.2 discusses some of the main policy measures affecting the producer prices of energy products (before consumption taxes) identified in the selected partner economies, providing the broader policy context for taxes on energy. Section 1.4.3 considers to what extent value-added tax (VAT) systems lead to changes in the relative prices of energy products compared to other consumption items through differentiated VAT rates in many of the 41 countries considered in this report. Whether or not energy taxes are set with environmental objectives in mind, the discussion provides useful insight into how pricing policies as a whole may or may not help to align prices with marginal costs, including environmental costs. Section 1.4.4 briefly discusses the current use of carbon trading mechanisms.

1.4.1. *Taxes on and tax expenditures for energy use*

Energy taxes increase the absolute and relative prices of energy products. They therefore impact energy use patterns, economic outcomes and the environment, and as argued above, specific taxes on energy can be very effective at integrating the environmental costs of energy use into usage decisions.

This analysis considers, on a systematic basis, taxes on the full spectrum of energy use in the 41 countries considered. The taxes covered in the detailed analysis are those levied on a physical measure of energy product consumed. They can be levied in a monetary amount per unit of fuel (per-unit taxes) or as a percentage of the sales price (*ad valorem* taxes). This report converts these tax rates into effective tax rates for each fuel based on, alternately, the energy and carbon content of each fuel. Taxes applying to a very broad range of goods (such as value added and retail sales taxes) are not included in the detailed analysis as they do not change relative prices. However, sometimes energy products are subject to a concessionary rate of VAT, which does affect relative prices. Section 1.4.2 provides insight into the extent of such VAT rate differentiation.

As explained in detail in Chapter 3 of Part I of this report, the pattern and level of taxes on energy use across the 41 countries vary considerably, both across countries and within individual countries for different uses and sources of energy. The way taxes affect the prices of energy use differs, and price signals are strongly heterogeneous across fuels and types of fuel use.

At the economy-wide level, there are large differences in the overall level of taxation across the 41 countries considered, both in energy and in CO_2 terms. The highest overall tax rates tend to be seen in countries which are members of the European Union, whose energy tax policy is significantly shaped by the 2003 EU Energy Tax Directive. Though far from being a homogenous group, the selected partner economies discussed in this report are among the jurisdictions with comparatively low average effective tax rates on energy use on an economy-wide basis, relative to the full group of 41 countries.

The economy-wide effective tax rates mask the diversity of tax rates on different fuels and users of fuels within individual countries and across the 41 countries as a whole. Transport energy is taxed more highly than heating and process energy and energy used for electricity generation. In addition, energy from oil products is taxed more heavily than energy from other sources. Several countries tax coal at very low rates, or do not tax coal at all. With the exception of Brazil, road transport is taxed at higher rates than other uses of energy. Of road use energy, diesel is taxed at lower rates in energy terms than gasoline in 39 of the countries considered, including in all selected partner economies except Brazil.

Regardless of the basis on which governments tax energy products, in practice they have often introduced exclusions or preferences to address potentially adverse impacts (real or perceived) of higher energy prices on particular groups of consumers or producers. It is increasingly recognised, however, that such preferences change relative prices in the economy in ways that have negative environmental impacts, lead to a loss of tax revenue, and create hurdles for increased use of alternative energy sources. Such tax expenditures are included in the analysis in as far as they are reported by the country concerned. Embedding the information on reported tax expenditures in the analysis of the taxation of energy use produces information complementary to the OECD's *Inventory of Estimated Budgetary Support and Tax Expenditures for Fossil Fuels* (OECD, 2015a, forthcoming), which focuses on the value of tax expenditures.

The need to phase out inefficient fossil fuel subsidies was recognised in the OECD's Declaration on Green Growth (OECD, 2009), which 42 countries have signed, and by the G20 leaders (G20, 2009). The OECDs *Inventory of Estimated Budgetary Support and Tax Expenditures for Fossil Fuels* shows that a significant amount of support in the countries analysed is provided through tax expenditures, including reductions in or exemptions from energy taxes.

A full assessment of tax expenditures requires broader consideration of the tax system of which they are a part. This report illustrates the value of the relief given under reported tax expenditures relating to taxes on energy consumption. In addition, it shows the broader context of these measures by showing the actual rate of tax as a result of the tax expenditure, the "normal" level of tax that would otherwise apply (the benchmark rate), and the rates of tax that apply to other products. This evidence-base allows cross-country comparison of effective tax rates and shows how taxes align with principles of environmental taxation. The report finds that there is considerable scope for better use of energy taxes to attain more environmentally-aware decisions on energy use.

1.4.2. *Main energy policies affecting producer prices of energy in the selected partner economies*

Governments can intervene in energy markets in a variety of ways with more or less direct effects on the producer prices of energy use, i.e. the prices before VAT and excise taxes. At one extreme, they can decide on prices in an *ad hoc* manner not guided by a transparent rule or process. At the other extreme, they can monitor and publish prices, indirectly affecting prices through better transparency and stronger competition. Between those extremes are different types of price regulation, price freezes, price bands and price ceilings, smoothing mechanisms, etc. Taxation or subsidisation at the production stage and the tax treatment of international trade also potentially affects domestic prices, as do quantity restrictions on international trade.

If there are strong differences among countries in the extent to which energy prices are dependent on non-tax policy instruments, and to the extent that these non-tax instruments do not aim to align prices with production costs, then direct comparisons of taxes to assess country differences in environmental and revenue characteristics of energy pricing policies can be misleading. A country where producer prices are in line with production costs and which does not levy taxes on energy use may send better price signals from an environmental point of view than a different country which does levy taxes on energy use but adds them to producer prices that are well below marginal production costs. Direct comparisons of tax profiles are useful as long as the impact of other energy pricing policies on price levels is not too strongly different.

It is beyond the scope of this report to provide a full and detailed description of all relevant measures, or to classify measures by their likely price impacts. Instead, Table 1 lists non-tax measures that affect end-user prices in the seven selected partner economies as identified in the country analyses undertaken for this report, by type of energy (coal, oil and oil products, natural gas and electricity). To provide the context in which these pricing policies apply, the country chapters refer briefly to the structure of the particular market for each broad fuel category. Inclusion of a measure in Table 1 is done on the basis of impacts on prices before VAT and end-user taxes, with no assessment of where the final incidence of the measure lies, or how broad the coverage of the measure is. As can be seen, a variety of pricing policies are being applied, most so in the sector of oil and refined petroleum products. The majority of measures listed in the table keep prices below market prices or production costs, through regulation, price freezes and direct price controls, perhaps reflecting an overall economic, social and political context that is conducive to maintaining prices below market levels even if awareness of drawbacks of such policies, and policy action to undo them, is growing.

1.4.3. *Differential VAT rates on energy products in the OECD and in selected partner economies*

Like the policies discussed in the previous section and the excise taxes discussed in detail in other parts of this report, value-added taxes (VAT) affect end-user prices of energy products in many jurisdictions, at least for those users that cannot claim back the input credits. As VAT applies to a very broad range – and ideally the broadest possible range – of goods and services in an economy, the tax is not specific to energy products and the relative price level of energy products and other goods and services is unchanged as long as the same VAT rate applies. However, if VAT rates are differentiated in a way that strongly affects the relative price of energy products, then VAT is *de facto* specific to such products and it should be considered when describing effective energy tax rates. Although not shown in the graphical profiles due to the difficulty of assessing and comparing the impact of differential VAT rates on energy products, particularly when differential rates also apply to other goods and services, this section investigates to what extent VAT rates for energy products differ from standard VAT rates. As in the previous section, the objective is to sketch to what extent VAT may interact with the objective of excise taxes, which is more precisely directed towards altering relative prices.

Table 1. Main energy policies affecting producer prices of energy

	Coal	Oil and oil products	Natural gas	Electricity	
				Generation	Electricity output
ARG	–	Dominant position of state-owned company in exploration and production (34% of the market) and refining (54%).	Dominant position of state-owned company in exploration and production (30% of the market), regional monopolies for transport and distribution.	Diversified market.	Private monopoly on transmission, regional monopolies on distribution.
		Direct price control at below market rates – *ad hoc* pricing for biodiesel, differentiated prices by company size. Retail price reductions in Jan. 2012, 2011 and Aug. 2010. **Price freeze** for LPG sold in bottles or cylinders in low-income areas. **Price monitoring** through publishing diesel and gasoline prices on government website. **Export taxes** for crude oil, biodiesel (both based on the difference between a national reference price and the international price) and natural gas (100%).	**Regulated prices at below-market rates – price freeze** since 2008, **differentiated** by consumer category and region.		**Regulated prices, differentiated** by user category and consumption except for some large users. **Earmarked fund** subsidises electricity tariffs. Electricity bill reduction for savings.
BRA	–	Dominant position of state-owned company in exploration and production (91%), refining (98%), distribution (40%), service stations (20%).	Dominant position of state-owned company in exploration, production, transmission (90%), distribution (70%) and imports (100%).	Dominant position of state-owned company (40%).	Publicly managed grid, distribution and retail mostly private.
		Direct price control through ownership, price freeze since 2006 (adjusted in 2012 and 2013), subsidy borne by supplier without reimbursement.	**Price regulation at above-market rates:** domestically produced natural gas is priced 1/3 higher than imports.	**Subsidised production inputs** for gas-fired power plants.	**Regulated prices** for non-industrial users.
		Extraction taxes and other taxes apply based on sales revenue of oil and gas extraction. **Import tax:** specific import tax is charged in addition to excise taxes. Special import tariff (14%) for biodiesel, programme provides imports tax exemptions for oil and oil products.	**Price regulation** at state level for downstream natural gas.	**Earmarked funds** subsidise off-grid diesel-, and coal-fired plants.	**Earmarked fund** subsidises tariffs for low-income users.
CHN	Primarily state-owned, relatively fragmented.	Dominated by state-owned companies in exploration and production, refining and distribution.	Dominated by state-owned companies in exploration and production, refining and distribution.	Dominated by state-owned companies.	Transmission and distribution managed by two state-owned companies.
	Price differentiation by user category and region at subnational level (coal prices can also differ between provinces because of a lack of transport infrastructure).	**Regulated prices – price ceilings** for gasoline, diesel fuel at wholesale and retail level and ethanol, possibility to introduce price ceiling for residential use of LPG at local level.	**Regulated prices at below-market rates** along the entire value chain – **price discrimination and cross-subsidisation** by consumer category and region.	**Regulated prices – *ad hoc* pricing** of coal inputs for power generation.	**Direct price control – *ad hoc* pricing, differentiated** for households.
		Price smoothing for gasoline and diesel according to a basket of crude every 10 work days, additional adjustments if international oil price exceeds certain levels, no regulation for crude oil.	**Price smoothing:** a pilot scheme bases gas-city prices on a weighted average of LPG and fuel oil prices in the Shanghai market with a discount, in 29 provinces.	**Subsidised production inputs** for Chongqing electricity producers.	Gradual elimination of preferential tariffs for large users, surcharges for heavy users.
		***Ad valorem* extraction tax** for oil, targeted reductions for some techniques and regions. **Export quotas** on oil products, temporary **export ban** on diesel in 2011.	**Extraction tax,** targeted reductions for some techniques and regions.		

Table 1. **Main energy policies affecting producer prices of energy** *(cont.)*

	Coal	Oil and oil products	Natural gas	Electricity	
				Generation	Electricity output
IDN	Production in private hands, relatively concentrated.	State-owned company controls 17% of production and operates nearly all refinery capacity, imports and supply.	State-owned company controls 13% of production and operates nearly all refinery capacity, imports and supply.	Dominated by state-owned company (85%).	Effective monopoly of state-owned company over distribution and retail.
	Regulated prices at below-market rates – *ad hoc* pricing.	**Direct price control at below market rates – *ad hoc* pricing** for crude oil, low grade diesel and gasoline, LPG and kerosene, cost borne by supplier with reimbursement, subsidised sales rationed using quota system. Multiple price increases in 2014 to reduce subsidies.	**Regulated prices at below-market rates – *ad hoc* pricing.**	**Subsidised production inputs** for coal fired-power plants.	**Regulated prices at below cost, uniform** for all consumers, cost borne by supplier with reimbursement.
	Domestic sales mandate: 1/5 of production.	**Domestic sales mandate** for oil. Subsidies are phased out for kerosene, LPG subsidy introduced for households and small businesses. Subsidies for biofuels.	**Domestic sales mandate.**		
IND	Dominated by state-owned companies across the entire value chain, deregulated in 2014.	Dominated by state-owned companies, particularly in refining and distribution.	Dominated by state-owned companies, particularly in refining and distribution.	Diversified.	State-owned company operates about 90% of the grid and transmits 50% of electricity.
	Direct price control through ownership. Subsidised sales are allocated between users.	**Direct price control at below-market rates – *ad hoc* pricing:** Kerosene and LPG price fix; **price differentiation**: Industrial kerosene is priced more than triple the household price. **Price monitoring** and indirect price fix through publishing diesel and gasoline prices on government website. Downstream oil companies compensated for under-recoveries related to transport of kerosene and LPG to remote areas, diesel, kerosene and LPG sales. Subsidised sales of kerosene and LPG rationed through public distribution system.	**Regulated prices, differentiated** for 1) state run companies; 2) joint venture fields and LNG imports; 3) LNG; 4) power and fertilizer producers; and 5) Northeast India.		**Regulated prices** on state level, some **differentiated** by user category depending on states.
RUS	Private and relatively fragmented.	Dominated by state-owned company in production and refining (40%), distribution (100%) and retail.	Dominated by state-owned company along the entire value chain.	State-owned companies control more than 60%.	Transmission under state control.
		Direct price control – *ad hoc* pricing: decrease gasoline prices in 2011 and provide diesel price discounts to farmers.	**Regulated prices at below-market rates** at wholesale and retail level for non-industrial consumers, **differentiated** by user category.		**Regulated prices, differentiated** by user category, region and technology.
		Regulated prices at below-market rates – price freeze for gasoline between Dec. 2011 and May 2012, temporary rationing of gasoline purchases in April 2011.	**Extraction tax with exemptions and reductions** for selected fields.		
		Extraction tax on crude oil, **exemptions and reductions** for selected fields.	**Export restriction:** export monopoly, monopoly lifted for LNG.		
		Export Tax on crude oil, gasoline, diesel and fuel oil, **exemptions and reductions** for selected fields.	**Export tax with exemptions and reductions** for selected fields.		

Table 1. **Main energy policies affecting producer prices of energy** *(cont.)*

	Coal	Oil and oil products	Natural gas	Electricity	
				Generation	Electricity output
ZAF	Privately owned. Prices set under long-term contracts between mining companies and major consumers (particularly Eskom); average price of coal is below current market prices.	Dominant position of state-owned company in production, other parts of the value chain are more diversified. **Price ceiling** for LPG and kerosene. **Price smoothing:** basic price for gasoline, kerosene and diesel set each month, difference to market price (+/-) covered by fund.	**Regulated prices at below-market rates – price ceiling** for piped based on weighted fossil fuel basket.	Dominant position of state-owned company.	**Regulated prices, differentiated** by consumer category and region. Free basic electricity policy.

Glossary of terms used in Table 1	
Direct price control	
Ownership	Direct ownership of companies at various levels along the supply chain, increases influence on pricing policy.
Ad hoc	Prices are adjusted at irregular intervals without a prescribed formula.
Price regulation or support	
Below-market	Regulating prices at below-market rates decreases user prices.
Above-market	Regulating prices at above-market rates ensures a certain profit margin for producers.
Uniform	Charge same price for all consumers.
Differentiated	Differentiated retail prices by user category or region (includes cross-subsidies).
Cash transfers, vouchers, earmarked funds	Decrease prices paid by all (uniform) or selected (differentiated) users.
Price freeze	Fix prices at the level of a certain date, sometimes for an unknown time period.
Price band	Regulated maximum and minimum prices, cannot be exceeded or undercut.
Price ceiling	Regulated maximum price, cannot be exceeded.
Price smoothing	Prices change according to a formula based on selected variables, sometimes automatic.
Price monitoring	Monitor prices, often via regularly publishing prices on government websites.
Subsidised production inputs	Conceptually similar to beneficial treatment under an extraction tax, subsidised production inputs decrease production costs.
Extraction tax	Levying tax on resource extraction increases the cost of this activity for producers. Beneficial treatment can be provided through tax exemptions or reductions.
Import tax	Levying tax on imports increases the price of imports relative to domestically produced goods.
Domestic sales mandate	Obliges companies to sell a proportion of their production in the domestic market, possibly at below-market prices.
Subsidised imports	Subsidising imports decreases their price on the domestic market.
Export tax and export restrictions	Creates a wedge between world and domestic prices, discouraging exports and decreasing domestic prices. Beneficial treatment can be provided for selected producers or fields through exemptions and reductions.

Table 2 lists the standard VAT rates in 40 of the 41 countries analysed[3] and the differential rates that apply to energy products in these countries.[4] Standard VAT rates differ substantially among these countries, ranging from 8% in Japan and Switzerland to 27% in Hungary. Out of the 40 countries considered, 21 are members of the European Union (EU), which requires member countries to apply a minimum standard VAT rate of 15%, while allowing one or two reduced rates of not less than 5% to apply to specified goods and services.[5] A number of derogations are in place that allow certain member states to continue charging differential VAT rates in addition to EU policy.

Table 2. **Differential VAT rates on energy products in selected partner economies and OECD countries, 2014**

	Standard VAT rate (%)	Differential VAT rate for energy (%)	Energy product subject to differentiated rate
Australia	10	–	
Austria	20%, specific regional rate: 19%	10	Firewood
Argentina	21	27	Natural gas Electricity (except public lighting)
		10.5	LPG Butane Propane
Belgium	21	12	Coal and solid fuel obtained from coal Lignite and agglomerated lignite Uncharred petroleum coke used as fuel
		6	Electricity (residential) Firewood
Brazil	1. VAT ("Imposto sobre Productos Industrializados", IPI) on industrial products: 5-300%	1. Exempts energy products	
	2. State sales tax (ICMS): 17-19%	2. Natural gas: 12%, Electricity: 25%	Electricity Natural gas
Canada	5% standard rate, specific regional rates: 13%, 14%, 15%.	–	Many provinces have sales tax reductions for energy products.
Chile	19	–	–
China	17	13	Natural gas LPG Biogas (res.) Coal, coal gas (res) Charcoal (res)
		4	Electricity by qualified hydro-electric generators
Czech Republic	21	15	Heating Firewood
Denmark	25	–	
Estonia	20	–	
Finland	24	–	
France	20	5.5	Natural gas Electricity District heating
		10	Firewood
Germany	19	7	Firewood
Greece	23	13	Natural gas Electricity District heating Firewood
Hungary	27	5	District heating
Iceland	25.5	–	Electricity and fuel oil used for the heating of houses and swimming pools
Israel	18	–	
Ireland	23	13.5	Energy for heating and light Natural gas Electricity Firewood Heating oil
India	Subnational sales taxes or VAT at 5-33%	4	Coal Crude oil Aviation fuel LPG for domestic use

Table 2. **Differential VAT rates on energy products in selected partner economies and OECD countries, 2014** *(cont.)*

	Standard VAT rate (%)	Differential VAT rate for energy (%)	Energy product subject to differentiated rate
Indonesia	10, government can vary the rate from 5 to 15%	Exempt	Electricity (res. < 6 600 Watt) LPG (3 kg cylinders) Crude oil, natural gas (unprocessed), coal (unprocessed)
Italy	22	10	Combustible gas for cooking Natural gas Electricity Firewood
Japan	8% (since April 2014)	–	
Korea	10	–	0% supply of mineral oil used for certain purposes in agriculture
Luxembourg	15	12	Solid mineral fuels Wood for fuel use (not for heating) Heat and air conditioning
		6	Natural gas Electricity Firewood LPG
Mexico	16	–	
Netherlands	21	–	
New Zealand	15	–	
Norway	25	–	Electric power and energy supplied by alternative sources in the counties of Finnmark, Troms and Nordland
Poland	23	8	Firewood
Portugal	23	13	Diesel (agriculture)
Russia	18	–	
Slovak Republic	20	–	
Slovenia	22	–	
South Africa	14	0 (zero rated)	Gasoline Diesel Kerosene
Spain	21	–	
Sweden	25	–	Aircraft fuel (kerosene)
Switzerland	8	–	
Turkey	18	–	
United Kingdom	20	5	Fuel and power for domestic and charity use

1. The *Imposto sobre Productos Industrializados* is a mix of a VAT and an excise tax levied on local and intrastate sales transactions of manufactured goods, at rates depending on their classification in the Harmonized Commodity Description and Coding System (HS) by the WCO. Subsequent manufacturers can take credit against IPI liability equal to the IPI paid by its suppliers.

Source: OECD, based on KPMG – VAT essentials (several dates) (selected partner economies); OECD (2014a) (OECD countries), Argentine Ministry of Finance (2013) (Argentina).

Twenty-one countries do not apply a differential VAT rate to energy products. Among the nineteen countries that do, with two exceptions, the rates are concessionary. Argentina levies VAT at a rate higher than the standard rate on some energy products, while Brazil's state sales tax levies a higher rate on electricity.[6] In the seventeen countries which have reduced or zero rates on energy products, the reduced rates are either set at approximately half the standard VAT rate or are substantially reduced to rates between 4% and 7%. Reduced or zero VAT rates are most frequently applied to electricity (13 countries), firewood (10), natural gas (9), LPG (4), district heating (4), heating oils (3), coal (3) and kerosene or aviation fuel (3), crude oil (3) diesel (2) and gasoline (1). Differential rates may apply only to specific users, e.g. all or some households, or small businesses.

One possible reason for reduced rates is policymakers' concern that low-income households spend a relatively larger share of their income on energy.[7] For example, countries with lower rates on residential heating fuels may place a relatively high weight on the ability of lower income households to afford heating fuels. To the extent this is true – and an analysis for 21 OECD countries shows it is less true for lifetime income as approximated by expenditures than for a snapshot of income at one point in time (OECD, 2014c) – it is not clear that VAT is the best instrument to address equity considerations, particularly in countries with broad, well-developed income taxation and social security systems, which may allow better targeting of support than the VAT system.

Among the countries considered, only Indonesia exempts residential electricity (and small LPG cylinders) from VAT. Unlike reduced or zero rates, which allow businesses to reclaim VAT paid on inputs, exemptions break the staged VAT payment system and introduce a cascading effect, as the non-deductible tax on inputs is embedded in the subsequent selling price and is not recoverable by taxpayers further down the supply chain. South Africa is the only country among those considered which has zero-rated some oil products (gasoline, diesel and kerosene).

Reducing VAT rates selectively for energy products counteracts the intention to increase the relative end-user prices of energy (for environmental or for revenue raising reasons). The effect of differential VAT rates on energy on the relative prices of energy is particularly pronounced when the differential rates apply only to energy products or to only a few additional products. By contrast, if the overall VAT system is characterised by strong differentiation of rates for broad sets of consumption items, then the impact of VAT differentiation on the relative prices of energy products is less easily established, but it will be weaker in general. OECD (2014a) provides details of differential VAT rates across all products. There is no apparent general pattern that countries with differentiated rates for energy products also allow differentiated rates for other goods. The degree of relative price differentiation for energy products through differentiated VAT rates is country-specific.

1.4.4. *Carbon trading mechanisms*

Like taxes, carbon emission trading schemes are a market-based measure that can be used to price carbon. Instead of taxing carbon emissions, governments can introduce an emission trading scheme by capping emissions and introducing tradable emission permits. Tradable carbon emission permits are similar to taxes in that they confront carbon emitters with a cost per unit of carbon emitted, equal to the price of a permit, and in that they allow the same flexibility in responses to reduce emissions. If the permits are auctioned instead of grandfathered or otherwise distributed at no cost, then public revenue is generated and the similarity between trading mechanisms and taxes is stronger. Trading schemes differ from taxes in that the price of tradable permits fluctuates with economic conditions (while the level of emissions does not change as long as the cap is not changed), whereas with taxes the cost of emitting a unit of carbon will not change with economic conditions (but the level of emissions will). However, accompanying measures can limit the range of price fluctuations under a trading scheme, in which case the difference between auctioned tradable permits and taxes becomes still smaller. Ultimately, the difference between the two approaches can become a matter of practical detail, and both approaches can efficiently price carbon.

The World Bank (2014a) estimates that currently about 12% of global greenhouse gas emissions are subject to an explicit carbon price through various mechanisms in around 40 countries and 20 subnational jurisdictions. In 2014, ETS covered more than 7% of global emissions. The 41 countries considered in this report have implemented a range of subnational, national and regional emission trading schemes (ETS), with others scheduled or planned for implementation. As observed in this short overview,[8] the design features of these schemes differ widely.

The EU ETS, operational since 2005, is the largest and longest-operating carbon emissions trading system. While historically most permits were grandfathered to emitters, the third trading phase will feature an increased proportion of auctioned permits. Prices continue to be relatively low, due to considerably slower economic activity after 2007 and also to interactions with other carbon abatement policies in EU member countries. Since the repeal of the Australian carbon pricing scheme in mid-2014, three national ETS exist in the countries considered. In New Zealand, trading started in 2008 (though emissions are uncapped). Switzerland introduced mandatory trading for energy-intensive firms in 2013, and a Korean ETS commenced in January 2015. This scheme started out with free allocation of emissions permits and auctioning will be slowly phased in starting in 2018. A national ETS is due to be introduced in China within the next few years, tentatively scheduled for 2016.

Several subnational emission trading schemes exist. The Californian cap-and-trade system started with voluntary participation in 2012 and obligatory compliance from 2013. The Californian system covers emissions from transport, agriculture, and households in addition to emissions from the industrial sector and power generation. Since January 2014, it has been linked to the ETS in Québec and there are efforts to to align emissions reduction policies, including carbon pricing, among several North American states in the context of the Pacific Coast Action Plan on Climate and Energy.

China has also introduced pilot ETS programs in seven cities and regions. The total emissions allocations of these pilot programs make China the second largest carbon market in the world after the EU ETS in terms of carbon emissions covered. While most pilot schemes use historical emissions or emissions intensity as a base for the free allocation of allowances, Guangdong is the first pilot scheme to have used auctioning as an allocation mechanism. This pilot scheme is also the largest among China's schemes and has the highest allowance prices (Munnings et al., 2014). The design of a national ETS in China, scheduled to be introduced between 2016 and 2020, will be based on the lessons learnt from the pilot schemes.

Other regional or partial trading schemes exist in Japan and in Brazil. In Japan, the Tokyo, Saitama and Kyoto trading schemes cover 8% of Japan's emissions, but only the first two mandate participation. While progress on the development of a national ETS in Brazil has stalled, emission trading has started among 22 companies, all members of the Businesses for Climate Platform.

2. Structure of the graphical profiles, methodology and data sources

The graphical profiles ("profiles") for each of the seven selected partner economies included in this report show the composition of energy use in each country covered and the effective rate of tax on various types of energy use. Both energy use and tax rates are shown, alternately, in terms of energy content and carbon content. The profiles also depict reported tax expenditures, showing both the actual tax rate and the benchmark rate against which the value of the preference is calculated.

This section provides an overview of the methodology, assumptions, and data sources underlying the profiles. Further details on these can be found in Annex A, or, where specific to a particular country, in the relevant country chapter.

2.1. *Tax base – energy use*

The horizontal axis of each graphical profile shows all final use of energy by businesses and individuals, including the net energy used in energy transmission and in the transformation of energy from one form to another (e.g., crude oil to gasoline, coal to electricity). Energy use has been grouped into three broad categories: transport, heating and process use, and electricity. These three categories have been further disaggregated for each country, generally reflecting the particular tax base of that country. The subcategories therefore differ between countries depending on the nature of the fuel, its user, or its use.

All forms of energy are converted into common units of energy (GigaJoules – GJ) and carbon emissions (tonnes of CO_2), using standard conversion factors. In the first graphical profile for each country, fuel quantities are expressed in terms of energy value (in GigaJoules). In the second graphical profile, the quantities of the various energy sources are expressed in terms of the carbon emissions associated with their use (in tonnes of CO_2).[9] The re-expression of tax bases in terms of carbon content permits a focus on the structure of taxation with respect to one purpose for which fuel can be taxed – to reflect the social cost of carbon emissions.

Electricity is different from most of the other energy types shown in that it is a secondary energy product which must be generated by use of some primary energy (e.g., coal, natural gas, nuclear power, and hydro). The electricity category of the graphical profiles therefore show the energy content or carbon emissions of the underlying primary fuels used to generate the electricity domestically rather than of the electricity itself.

Data on energy use is taken from the *Extended World Energy Balances* (IEA, 2011).

2.2. *Tax rates and tax expenditures*

On the vertical axis, the graphical profiles show the tax rates and related tax expenditures that apply to energy use as at 1 April 2012 (except for Australia and Brazil, which are shown as at 1 July 2012, and South Africa, which is as at 4 April 2012). The taxes covered are those levied on a physical measure of energy product consumed, whether quoted in a monetary amount per unit of fuel (per-unit taxes), or as a percentage of the sales price (*ad valorem* taxes). In contrast to the 2013 report *Taxing Energy Use – A Graphical Analysis* (OECD, 2013b), *ad valorem* taxes (i.e. not including VAT) and related tax expenditures, set by reference to the value of products are included in the graphical profiles in this report. This is because in several of the selected partner economies, non-VAT taxes on energy products levied as a percentage of the sales price are more commonly used (for example, in Argentina, India and Indonesia). Where taxes on energy are quoted as a percentage of the sales price, price information was used to translate *ad valorem* rates into per-unit rates. Converting *ad valorem* taxes into per-unit taxes allows the calculation of effective tax rates on energy and carbon terms across different bases, but the calculated unit taxes are contingent upon observed prices.

Taxes that apply to a very broad range of goods (such as value added and retail sales taxes) are not included in the graphical profiles on the basis that since they apply equally to a wide range of goods, they do not change relative prices. However, where an energy product is subject, for example, to a concessionary rate of VAT, the concession would affect relative prices. In order to gauge to what extent VAT rate differentiation takes place, VAT and concessionary VAT rates on energy products are discussed in Section 1.4.2. Also excluded are taxes that that may be related to energy use but that are not imposed directly on the energy product (such as vehicle taxes, road user charges or billing charges and taxes on emissions such as NO_X and SO_X) and those which do not have a fixed relationship to fuel volume.

Taxing Energy Use – A Graphical Analysis (OECD, 2013b), as well as this report, consider the impact of energy taxes on the user price of energy from a common base of zero. When taxes are added to producer prices of energy which align broadly with production costs, the end user price aligns with social costs if the tax approximates external costs. However, some of the countries analysed in both this report and in *Taxing Energy Use – A Graphical Analysis* (OECD, 2013b) apply price support measures that keep producer prices below production costs. Countries can also apply production taxes, royalties and other levies on the extraction or harnessing of energy resources, or levy export taxes, all of which may affect producer prices.

Non-tax pricing policies are not shown in the graphical profiles since it was not possible to obtain detailed information on relevant prices for all products and users, as well as on those prices which would apply in the absence of these policies. Furthermore, production and export taxes are usually not directly imposed on the end-users of energy products, and the rate to which they are passed through can vary largely.

Tax rates on the use of energy are re-calculated as effective tax rates per GigaJoule of energy (in the first graphical profile for each country) and per tonne of CO_2 emissions (in the second graphical profile). Tax rates are shown in local currency on the left-hand axis of the graphical profile, and in Euros on the right-hand axis (converted by reference to the average market exchange rates over the 12 months ending August 2012). The tax rate applying to each fuel is shown on the graph as a shaded bar across the portion of energy use or carbon emissions (the tax base) to which the particular rate applies. The shaded rectangle beneath this bar is an approximation of the revenue raised by the tax – the rate multiplied by the base.

Taxes levied on electricity consumption have been shown as effective taxes on the fuels used to generate the electricity. In cases where a common nominal tax rate is applied to all electricity consumption, the effective tax rate on each underlying energy source (e.g., coal, natural gas, hydro) used to generate the electricity is shown. In cases where different rates of nominal electricity tax apply to consumption in different sectors (e.g., residential, commercial, industrial), for each sector, the effective tax rate shown is that of the "average" basket of fuels used to generate electricity in the country. Notably, when there is a general tax on electricity consumption that applies regardless of the generation source, and if carbon energy is a small proportion of the generation mix, the effective tax rate on carbon thus calculated will be high. A tax on electricity consumption that does not distinguish between electricity from carbon sources and electricity from non-carbon sources does not send an effective price signal about the use of carbon.

The graphical profiles also show tax rebates, credits and other tax expenditures that are reported by the country concerned. In the graphical profiles, the area of the light grey shaded rectangles is an estimation of the amount of the tax expenditure. In addition, however, the top of this rectangle is the benchmark or "normal" level of tax from which the measure is a departure while the bottom of the rectangle is the net level of tax that applies as a result of the concession. In the case of tax expenditures from *ad valorem* taxes, the top and bottom of the rectangle show the *ad valorem* benchmark tax treatment and the net level of *ad valorem* tax treatment of the energy product, respectively, converted into a per-unit tax rate.

In this respect, the graphical profiles are a complement to analysis that focuses on the value of tax expenditures, such as the OECD's *Inventory of Estimated Budgetary Support and Tax Expenditures for Fossil Fuels* (OECD, 2013a, 2015, forthcoming). By showing tax expenditures in context, the graphical profiles can facilitate discussion about appropriate tax benchmarks for different fuels, uses and users.

Given the economy-wide scale of the profiles, they do not show certain small details of tax bases, rates and preferences. Where multiple energy taxes or tax components apply to the same base, they have been aggregated. Where important, energy taxes at the sub-national level are indicated for an illustrative group of states or provinces.

Some countries price carbon emissions for some sectors through emission permit trading. The graphical profiles note the interaction of tax systems with emission trading systems (ETS) where applicable, but does not include the price provided by these schemes in the effective tax rates shown. For example, the graphical profiles of countries participating in the European ETS shown in *Taxing Energy Use – A Graphical Analysis* (2013b), distinguish subcategories of energy use which are entirely or partially covered by the ETS. Of the selected partner economies, China has implemented emissions trading on a subnational basis. A description of the seven pilot schemes can be found in the country chapter on China in Part II of this report. However, the trading systems are not shown in the graphical profiles for China as they were not in operation at the time to which these profiles apply.

The underlying data shown in the graphical profiles in this publication, as well as those presented in the previous publication of *Taxing Energy Use* (OECD, 2013b) is used to compile a comparable database of effective tax rates on energy, and energy use for all countries. This database is highly disaggregated by fuel and user and is used to calculate average effective tax rates across a wide range of fuels and users, weighted by the amount of energy subject to each rate. It therefore provides a systematic and comparable basis of analysis across the group of 41 countries. In order to provide some insight into the variety of policy measures that affect producer prices of energy, a summary of pricing measures applied in the selected partner economies is presented alongside their graphical profiles and the main policies affecting producer prices were discussed in Section 1.4.

Information on taxes has been taken from country-specific sources, from the OECD/EEA database on instruments used for environmental policy (OECD, 2014b), from Kojima (2013), and in some cases from country analysis briefs by the Energy Information

Administration (EIA, several years). Tax expenditure information is primarily from OECD (2013a). For Argentina, tax expenditures were retrieved from the Argentinean Ministry of Finance's report on tax expenditures (Argentina Ministry of Finance, 2012).

Further information on the methodology applied in this report can be found in Annex A.

3. Energy use and taxation: Results from the analysis

The graphical profiles presented in the second part of this report provide a number of insights into the taxation of energy use in Argentina, Brazil, China, India, Indonesia, Russia and South Africa. This section uses the data presented in the seven graphical profiles for these countries, together with the information presented in the graphical profiles for OECD countries in *Taxing Energy Use – A Graphical Analysis* (OECD, 2013b) to analyse energy use and taxation patterns across all 41 countries. Together, these countries account for just over 80% of the world's energy use and for 84% of the world's carbon emissions from energy use.

This section considers energy use patterns in each country before examining energy use and taxation in each of the three broad categories of energy use shown in the graphical profiles in more detail: transport, heating and process use and electricity. Finally, the section discusses the economy-wide levels of energy taxation in each country and puts the effective tax rates on energy use into a broader framework, considering energy use and emissions per capita and per unit of GDP, drawing on the context provided in Section 1.3.

The graphical profiles demonstrate that the energy landscape across these 41 countries is quite diverse both in use and in tax patterns. The uses and sources of energy are vastly different between countries. Likewise, countries tax energy in many ways, with variations in tax bases, tax levels, rebates and reported tax expenditures. The patterns and levels of energy taxation vary considerably both across countries and within individual countries for different uses and sources of energy. Countries therefore differ strongly in how taxes affect the prices of energy use, sending very different price signals in respect of different fuels and fuel uses.

3.1. General trends in energy taxation across countries

3.1.1. Uses and sources of energy

Across the 41 countries considered, patterns of energy use vary significantly, both in terms of the sources and uses of energy. Data on energy use and tax rates in these countries is taken from the detailed graphical profiles of energy use and taxation in the selected partner economies shown in this report and in the OECD countries shown in OECD (2013b).

The graphical profiles show energy use divided into three broad categories: transport energy, energy used for heating and process purposes and energy used in electricity generation. The proportion of energy used in each of these categories varies considerably between countries. When considered as a share of carbon emissions from energy use, the relative contribution of each of these three categories will also be dependent on the respective carbon intensity of the fuels used in each category. For example, in countries that rely more strongly on renewable sources of electricity generation, the share of electricity generation energy in total carbon emissions, all else being equal, will be smaller than its respective share in total energy use. Differences in the fuel mix of each country mean that for some countries the proportions of each category differ considerably in terms of energy use as compared to carbon emissions from energy use.

Figure 2 shows the proportion of energy use (left panel) and CO_2 emissions (right panel) in each of the three main categories for each of the 41 countries considered. Of the three categories of energy use, the proportion of transport to total energy use and to emissions from energy use ranges from 6% of total energy use in Iceland (but 43% of carbon emissions from energy, due to the high volume of renewables used in electricity generation), to 65% of energy use (67% of carbon emissions from energy) in Luxembourg. However, both of these countries can be seen as outliers. Luxembourg has a very high level of transport energy both relative to other energy use and relative to other countries' shares of energy use (for example, Slovenia, the second highest, uses only 33% of its energy for transport purposes, even though it also attracts fuel tourism

Figure 2. **Composition of energy use and of CO_2 emissions by use**

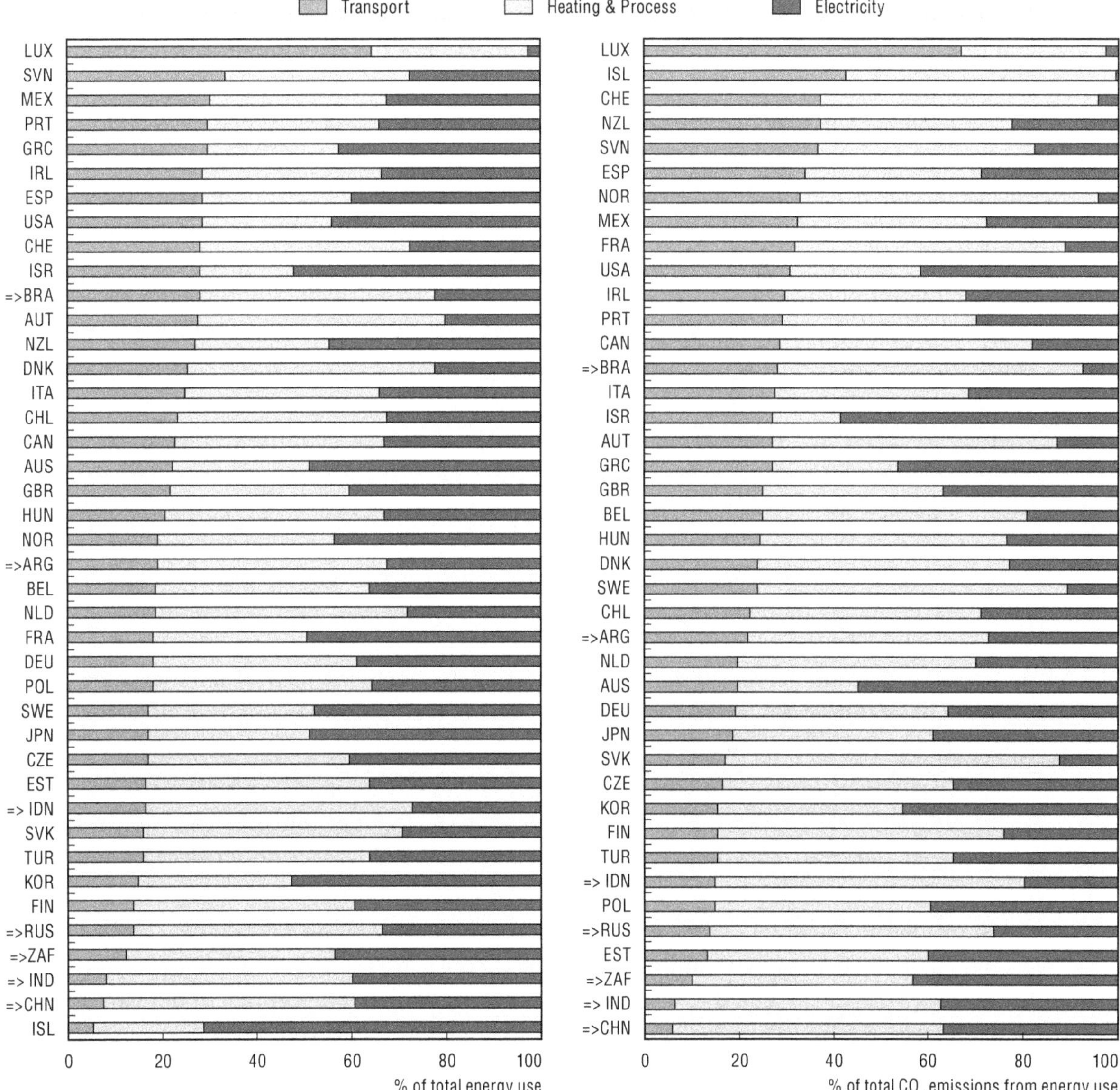

Note: Composition of energy use (left panel) and of CO_2 emissions (right panel) by use.

Source: OECD calculations based on energy use data for 2009 from IEA (2014), *IEA World Energy Statistics and Balances* (database), Doi: http://dx.doi.org/10.1787/data-00513-en.

StatLink http://dx.doi.org/10.1787/888933205556

due to comparatively lower transport rates), due to the high volume of motor fuel sales to non-residents. Iceland's small share of transport energy is due to the extremely high amount of energy used in electricity generation (71%), which is well above the share of electricity generation energy in total energy in other countries.

Excluding these outliers, China and India have the lowest transport shares among the 41 countries considered, at 8% of total energy use each (6% of carbon emissions from energy) and South Africa and Russia also have comparatively low shares of transport energy relative to other countries, at 13% and 14% respectively (10% and 14% of carbon emissions from energy). The unweighted country average share of transport energy to total energy is 22% and the average level of carbon emissions from transport energy is 25% of total energy emissions, across all 41 countries considered. With the exception of Brazil (28% of energy and carbon emissions from energy), the selected partner economies all have comparatively low shares of energy use in transport, although in China, Indonesia and India the share of transport energy to total energy use has grown rapidly since 1990 (IEA, 2011).

The share of energy used in the heating and process category ranges from 20% of energy use (14% of CO_2 emissions) in Israel to 56% in Indonesia (37% of CO_2 emissions) and all but eight countries use between 35% and 50% of energy for heating and process purposes. Heating and process use accounts for 41% of total energy use on an unweighted country average and for 48% of total carbon emissions from energy. The selected partner economies typically use larger than average shares of energy use for heating and process purposes, equating to over 50% of total energy use in all selected partner economies except South Africa (44%).

The proportion of energy used in electricity generation ranges from 20% of total energy in Austria to 53% in Korea, with a simple unweighted country average of 37%, again excluding Iceland and Luxembourg. Within the selected partner economies, the share of energy use in electricity generation is lower, varying from 22% in Brazil to 43% in South Africa. However, when considered in carbon terms, the picture is significantly different due to the high diversity in energy sources for electricity generation in different countries. Excluding outliers, Switzerland has the lowest share of carbon emissions from energy used in electricity generation (4%); while Israel has the highest at 58%. Countries that use significant shares of renewable or nuclear electricity generation have low proportions of carbon emissions from energy use in electricity generation, notably Brazil, France, Norway and Switzerland.

The sources of energy also vary substantially across countries. Figure 3 disaggregates energy use (left panel) and CO_2 emissions from energy use (right panel) into five major fuel groups: coal and peat, oil products, natural gas, biomass and waste and renewables and nuclear.

Considering the whole range of energy used in each economy, oil products are the primary source of energy, accounting for 34% of energy use and 39% of carbon emissions from energy use (unweighted country averages). However, the proportions in individual countries range from 11% in Iceland to 72% in Luxembourg, again reflecting the unusual characteristics of energy use in these countries. Even excluding these outliers, there is still a considerable range, from 14% in China and 16% in South Africa to 57% in Chile. Oil products are the largest source of energy in 21 countries. The share of coal in the economy-wide energy mix also varies considerably between countries, accounting for 1%

of all energy in Switzerland to 70% in South Africa. Coal is the dominant source of energy in 9 countries: Australia, China, Estonia (oil shale), India, Israel, Korea, Poland, Turkey and South Africa. On an unweighted country basis, natural gas accounts on average for 21% of energy, although the range between countries is more limited than that of coal or oil. Only Argentina, the Netherlands and Russia derive more than 50% of total energy from natural gas and it is also the dominant fuel (although with a share of less than 50%) in Italy, Hungary, the Slovak Republic and the United Kingdom. For four countries, renewables or nuclear energy is the largest single source of energy (France, Iceland, Norway and Sweden).

Figure 3. **Composition of energy use and of CO_2 emissions by fuel**

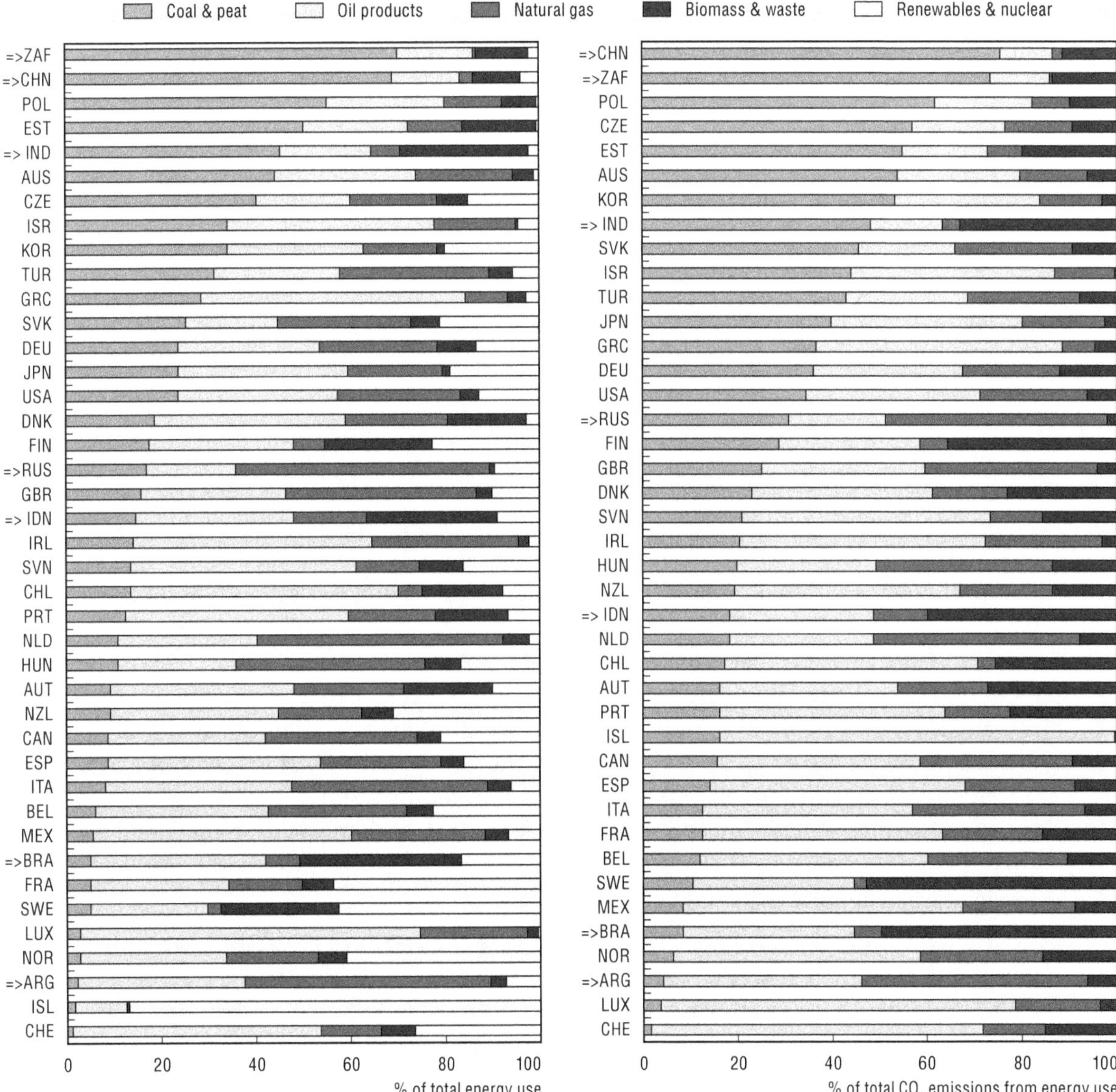

Note: Composition of energy use (left panel) and of CO_2 emissions (right panel) by fuel.

Source: OECD calculations based on energy use data for 2009 from IEA (2014), *IEA World Energy Statistics and Balances* (database), Doi: http://dx.doi.org/10.1787/data-00513-en.

StatLink http://dx.doi.org/10.1787/888933205562

Figure 4 shows the amount of energy use and the carbon intensity of fuel use across all 41 countries. The horizontal axis shows the cumulative proportion of energy use from each fuel, ranked from those with the lowest carbon intensity to the highest. The vertical axis shows the amount of CO_2 emitted when a TJ of each fuel is consumed.

Figure 4. **Carbon intensity and amount of energy use by different fuels across all 41 countries**

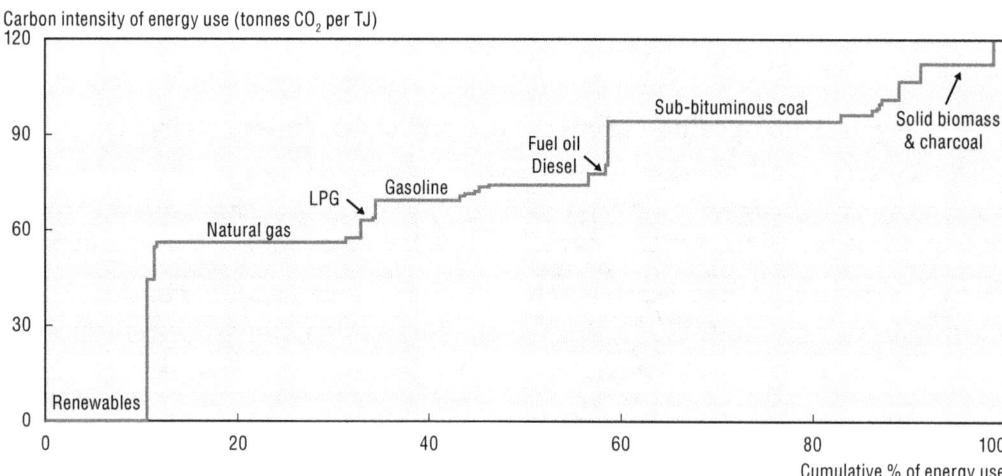

Source: OECD calculations based on energy use data for 2009 from IEA (2014), *IEA World Energy Statistics and Balances* (database), Doi: http://dx.doi.org/10.1787/data-00513-en. Conversion factors for different fuels may vary slightly between countries.

StatLink http://dx.doi.org/10.1787/888933205573

Renewable sources of energy (other than biofuels) and nuclear account for 11% of all energy use across the 41 countries considered and have no carbon emissions per unit of energy use. Natural gas, which represents 20% of total energy use by these countries, has around 56 tonnes of CO_2 per TJ. Oil products range from 63-75 tonnes of CO_2 per TJ, with the most commonly used oil products, gasoline (9% of all energy use) and diesel (10%), ranging from 69 to 74 tonnes of CO_2 per TJ. The most heavily-used coal products have a carbon intensity which ranges from 80 to100 tonnes of CO_2 per TJ and the most commonly used form of coal, sub-bituminous coal, has a carbon intensity of 96.1 tonnes of CO_2 per TJ. Although not commonly used, certain coal gases have higher carbon intensities, of up to 260 tonnes of CO_2 per TJ. Solid biomass and charcoal (7% of energy use, with the vast majority being derived from biomass), has a carbon intensity of around 110 tonnes of CO_2 per TJ. Other than solid biomass and charcoal, combustibles and waste energy products are not clearly identifiable on Figure 4, due to their small amounts of use, and have widely varying carbon intensities per unit of energy.

However, within each of the different categories of energy use lie vastly different fuel mixes. The amount of energy use from different sources by different users is summarised for all countries, on a weighted average basis, in Figure 5. When all energy use in the 41 countries is considered as a whole, transport energy is dominated by oil products (94% of all transport energy in the 41 countries considered), particularly by gasoline and diesel; consequently, transport uses the greatest share of oil products (63% of all oil products), relative to the other use categories. For heating and process energy, the fuel mix is more diverse, with coal and natural gas each accounting for 31% of heating and process energy and oil products (21%) and combustibles and waste (17%) are also significant sources of heating and process energy across the 41 countries considered. In the electricity category, coal is the primary source of energy used to generate electricity (53% of all energy used to generate electricity across all 41 countries considered), with renewables and nuclear energy accounting for just over 27% of energy used in electricity generation. The electricity category uses the highest proportion of coal (62%) of any use category and also uses almost all renewable and nuclear energy (98%).

Figure 5. **Sources and uses of energy for total energy use across all countries (weighted average basis)**

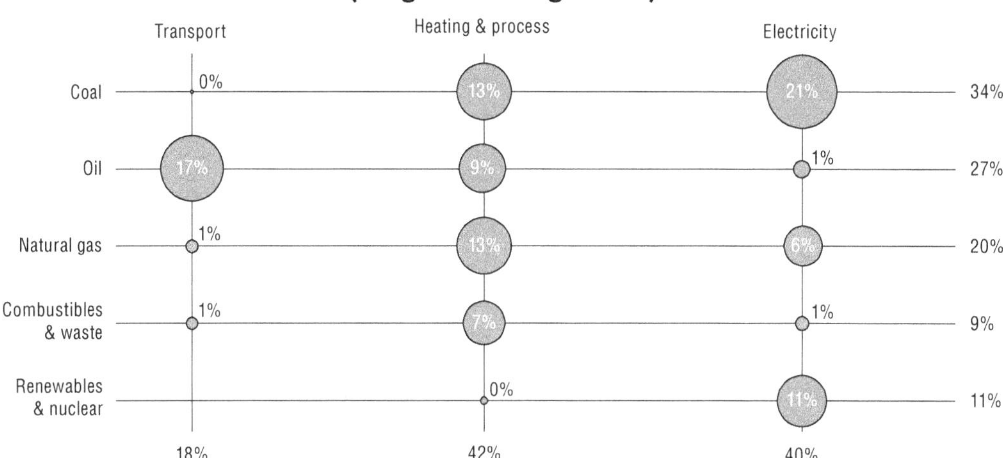

Source: OECD calculations based on energy use data for 2009 from IEA (2014), *IEA World Energy Statistics and Balances* (database), Doi: http://dx.doi.org/10.1787/data-00513-en. Percentages indicated in circles may not sum to the totals indicated due to rounding.

StatLink http://dx.doi.org/10.1787/888933205582

Each of the different fuel groups has different characteristics in terms of carbon emissions and local air pollutants associated with a GigaJoule of energy. The carbon intensity of energy use varies with each fuel; while these factors may vary slightly between countries, they are roughly consistent across fuels. Broadly speaking, coal has the highest level of carbon emissions per unit of energy, with different coal products ranging from 80 to 260 tonnes of CO_2 per TJ. Oil products have a slightly lower carbon intensity per unit of energy, ranging from 63-75 tonnes of CO_2 per TJ. Natural gas has a lower carbon intensity, at around 55 tonnes of CO_2 per TJ. The carbon intensity of renewables and waste differs markedly by source: some forms of waste or biomass have carbon intensities in the region of those of coal, whereas others, particularly for gaseous biomass, are lower than those for natural gas. Finally, renewables from many sources, such as hydro, solar, geothermal or wind, as well as nuclear, have no carbon emissions per unit of energy.

As each of the three categories of energy use have different fuel mixes, their relative carbon intensities, defined as the amount of carbon emitted for each unit of energy consumed in each category, differ. As energy use in transport is dominated by oil products, the carbon intensity of transport energy in most countries is very similar to that of oil products used in this category: 36 countries have a carbon intensity of energy used in transport of between 70 and 73 tonnes of CO_2 per TJ of energy use and the range is only from a minimum carbon intensity of 66 tonnes of CO_2 per TJ in Russia, which uses a comparatively high share of natural gas in transport, to 73.2 tonnes per TJ in Belgium.

Heating and process use energy is more diverse, ranging from coal to natural gas, with varying proportions of each being used in each country. The carbon intensity of heating and process energy in each country is therefore more varied, ranging from 53 tonnes of

CO_2 per TJ in Israel, which uses the highest proportion of renewables in this category, to 100 tonnes of CO_2 per TJ in China, where heating and process energy is primarily derived from coal. The use of coal as a source of heating and process energy, particularly in larger economies, together with the comparatively small share of renewable energy sources in this category, means that the heating and process category has, on average, the highest carbon intensity of energy among the three use categories.

Across the countries considered, the electricity category has the most diverse range of carbon intensities per unit of energy, as the range of fuels to generate electricity on a per-country basis are the most diverse of any category. Several countries, including Australia, China, Estonia (oil shale) and South Africa use high proportions of coal in electricity generation; others, such as Luxembourg and Russia use natural gas as a primary source of electricity generation; and several, use high proportions of renewables (Brazil, Iceland, Norway and Switzerland) or nuclear (France). This results in highly varied carbon intensities of energy for electricity generation for different countries. The distribution of the carbon intensity of energy in each country for each of these categories is shown in Figure 6, where the height of the bars (and the labels) shows the number of countries at the different levels of carbon intensity per unit of energy.

Figure 6. **Distribution of the carbon intensity of energy use in each category by country (number of countries in each range in brackets)**

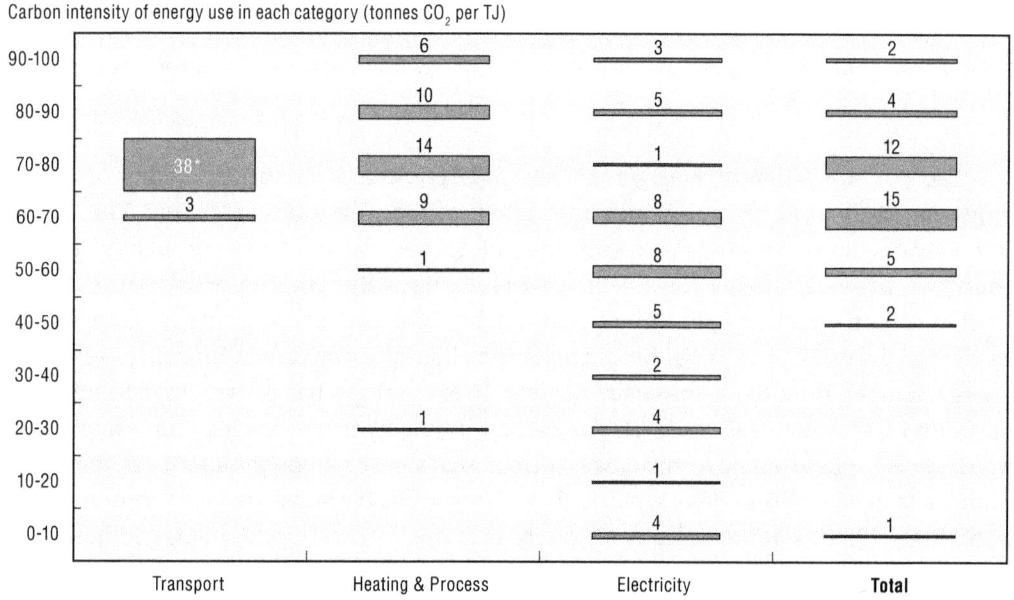

* In the transport column, at the level indicated as 70-80 tonnes CO_2 per TJ, all of the 38 countries are within a range of 70-73.2 tonnes of CO_2 per TJ.

Source: OECD calculations based on energy use data for 2009 from IEA (2014), *IEA World Energy Statistics and Balances* (database), Doi: http://dx.doi.org/10.1787/data-00513-en.

StatLink http://dx.doi.org/10.1787/888933205591

3.1.2. *Tax rates on energy across the 41 countries*

Tables 3 and 4 show the effective tax rates on the average unit of energy (Table 3) and CO_2 emissions from energy (Table 4), disaggregated by major fuel types and fuel use categories. Across the 41 countries considered, the weighted average tax rate on a unit of energy is EUR 1.1 per GJ and EUR 14.8 per tonne of CO_2. However, the weighted average tax rates vary for energy from different fuels and for different users. On average, transport energy is taxed more highly than that derived from other fuels; at EUR 5 per GJ and EUR 70.1 per tonne of CO_2 (weighted averages). Heating and process energy and energy used for electricity generation are taxed at similar rates in both energy and carbon terms (at a weighted average rate of EUR 0.3 per GJ and EUR 3.1 and 3.4 per tonne of CO_2, respectively).

Table 3. **Weighted average effective tax rates on energy by fuel type and use (EUR per GJ)**

		Oil products	Coal and peat	Natural gas	Biofuels and waste	Renewables and nuclear	All fuels
	% of base	27%	34%	20%	9%	11%	100%
Transport use	18	5.20	0.00	0.12	3.74	0.00	4.96
Heating and process use	42	0.82	0.05	0.21	0.00	0.00	0.26
Electricity production	40	0.50	0.13	0.43	0.65	0.38	0.27
Total use	100	3.52	0.10	0.28	0.30	0.38	1.11

Source: OECD calculations for selected partner economies; OECD (2013b), *Taxing Energy Use – A Graphical Analysis*, Doi: *http://dx.doi.org/10.1787/9789264183933-en*, for all other countries. Tax rates are as of 1 April 2012 (except 1 July 2012 for AUS and BRA and 4 April 2012 for ZAF); energy use data is for 2009 from IEA (2014), IEA *World Energy Statistics and Balances* (database), Doi: *http://dx.doi.org/10.1787/data-00513-en*. Figures for CAN, IND and USA include only federal taxes.

StatLink http://dx.doi.org/10.1787/888933205946

Table 4. **Weighted average effective tax rates on CO_2 from energy use by fuel type and use (EUR per tonne CO_2)**

		Oil products	Coal and peat	Natural gas	Biofuels and waste	All fuels
	% of base	26%	46%	15%	13%	100%
Transport use	17	72.89	0.00	2.13	51.84	70.05
Heating and process use	48	11.60	0.48	3.75	0.01	3.07
Electricity production	35	6.87	2.31	5.85	16.36	3.37
Total use	100	49.32	1.58	4.37	3.61	14.78

Source: OECD calculations for selected partner economies; OECD (2013b), *Taxing Energy Use – A Graphical Analysis*, Doi: *http://dx.doi.org/10.1787/9789264183933-en*, for all other countries. Tax rates are as of 1 April 2012 (except 1 July 2012 for AUS and BRA and 4 April 2012 for ZAF); energy use data is for 2009 from IEA (2014), IEA *World Energy Statistics and Balances* (database), Doi: *http://dx.doi.org/10.1787/data-00513-en*. Figures for CAN, IND and USA include only federal taxes.

StatLink http://dx.doi.org/10.1787/888933205955

The higher tax rates on transport energy may be explained by the broader range of policy goals that governments address in this category, as well as by the use of transport energy taxes for revenue purposes. Although the contribution of energy to carbon emissions does not vary depending on its use, transport use of fuel is indirectly tied to other externalities, such as local air pollution, congestion, accidents and noise, and taxes on transport energy may be used as an indirect means of internalising these externalities. In addition, a number of countries formally or informally earmark revenue from transport energy taxes to fund infrastructure.

There is also an underlying diversity in how different fuels are taxed. Energy from oil products is taxed more heavily than energy from other sources. While part of this is explained by the higher tax rates applying to transport energy, which is almost exclusively derived from oil products, the same pattern of higher taxation of oil products holds within each of the other categories, with the exception of biofuels and waste used for electricity generation. Natural gas, biofuels and waste are taxed at lower rates, at EUR 0.3 per GJ and EUR 4.4 and 3.6 per tonne of CO_2. Energy from coal is taxed at the lowest rates in both energy and carbon terms, at EUR 0.1 per GJ and EUR 1.5 per tonne of CO_2. On average, coal used in the heating and process category is taxed at the lowest rates of all fuel and use combinations – at less than EUR 0.0004 per GJ and EUR 0.5 per tonne of CO_2 – while coal used in electricity generation is taxed at a slightly higher rate. The amount of coal used in transport, while taxed at very low rates, is negligible.

Some of these patterns are also observed in the selected partner economies. Transport energy is taxed more heavily than other forms of energy in all selected partner economies except Brazil, where between the suspension of the CIDE tax in June 2012 and its re-instatement in February 2015, transport fuels were *de facto* untaxed. Similarly, oil products, particularly those used in transport, are taxed at higher rates in all countries (except Brazil). Natural gas and coal are taxed at lower rates and are more frequently untaxed. Full results for these countries are set out in Annex B. Full results for OECD countries can be found in Annex B of OECD (2013b).

Figure 7 presents a graphical profile of average effective tax rates and energy use across the main fuel categories for total energy use across the 41 countries considered. Figure 8 shows the same information in terms of the carbon emissions from energy use in these countries. These figures show the higher tax rates applied to fuels in the transport category, particularly oil products, which account for almost all energy use and carbon emissions from this category. The average effective tax rates applied in the heating and process use and electricity categories are considerably lower, and there is variation within these categories for different fuels. In both the heating and process category and the electricity category, coal is taxed at lower rates than other fossil fuel sources of energy, with oil products facing the highest rate. The lower tax rate on coal is particularly pronounced in the graphical profile shown in carbon terms, due to the higher carbon intensity of coal. From an environmental perspective, the lower tax rates on coal do not reflect the social costs associated with its use. It has a greater carbon content, per unit of energy, and is also associated with higher levels of air pollutants (particularly sulphur dioxides and particulate matter, as well as smaller contributions from nitrogen oxides), although the level of air pollutants emitted per unit of energy varies with the location and height of emissions, as well as any control technologies used. However, in many countries the mortality and health costs associated with air pollution are significantly higher per TeraJoule of coal than per TeraJoule of natural gas, gasoline or diesel (Parry, 2014).

The contrast between the relatively high taxation in the transport category and the relatively low taxation in the heating and process and electricity categories is very evident from both graphical profiles. Equally evident in the graphical profiles is the variation in tax rates on different fuels within each of the three use categories. Sections 3.2 to 3.4 of this chapter examine these differences in treatment within each of the three broad use categories: transport, heating and process use and electricity generation.

3.2. *Taxation of energy used in transport*

3.2.1. *Users and sources of transport energy*

The transport category includes road transport and rail, marine and domestic air transport. Across the 41 countries considered, transport accounts for around one-fifth to one-quarter of energy use and a slightly higher proportion of carbon emissions from energy use, although in several countries, transport energy accounts for a lower proportion of both energy use and emissions. Road transport is the primary user of energy in this category in all countries, accounting for 87% of total energy use. However, as a result of the substantial tax rates highlighted above, it generates the vast majority of revenue from energy taxes in almost all countries.

Figure 7. **Graphical profile of energy use and taxation across all energy use in the 41 countries (weighted average basis)**

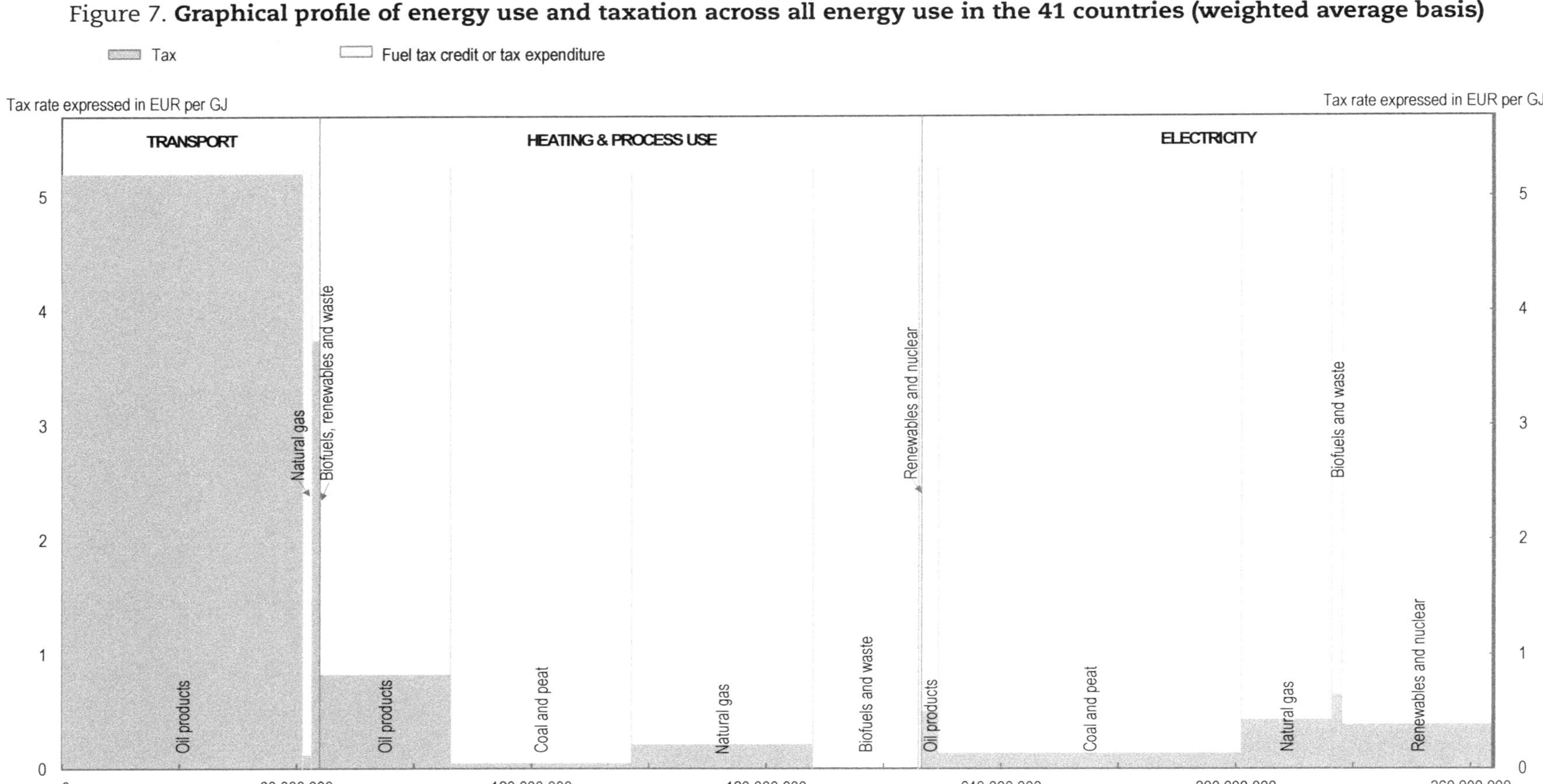

Source: OECD calculations for selected partner economies; OECD (2013b), *Taxing Energy Use – A Graphical Analysis*, Doi: http://dx.doi.org/10.1787/9789264183933-en, for all other countries. Tax rates are as of 1 April 2012 (except 1 July 2012 for AUS and BRA and 4 April 2012 for ZAF); energy use data is for 2009 from IEA (2014), *IEA World Energy Statistics and Balances* (database), Doi: http://dx.doi.org/10.1787/data-00513-en. Figures for CAN, IND and USA include only federal taxes.

StatLink http://dx.doi.org/10.1787/888933205609

Figure 8. **Graphical profile of energy use and taxation across all carbon emissions from energy use in the 41 countries considered (weighted average basis)**

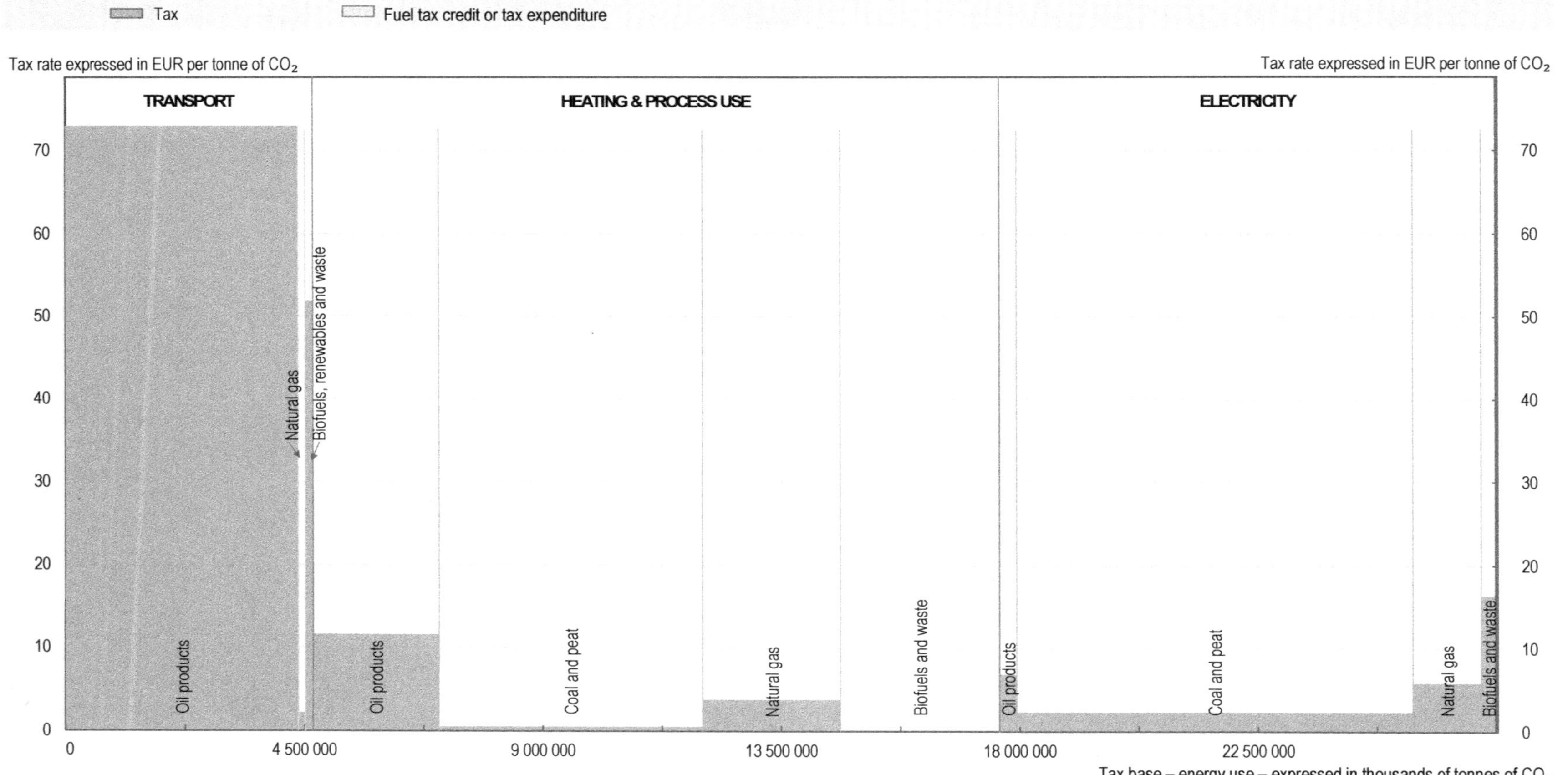

Source: OECD calculations for selected partner economies; OECD (2013b), *Taxing Energy Use – A Graphical Analysis*, Doi: http://dx.doi.org/10.1787/9789264183933-en, for all other countries. Tax rates are as of 1 April 2012 (except 1 July 2012 for AUS and BRA and 4 April 2012 for ZAF); energy use data is for 2009 from IEA (2014), *IEA World Energy Statistics and Balances* (database), Doi: http://dx.doi.org/10.1787/data-00513-en. Figures for CAN, IND and USA include only federal taxes.

StatLink http://dx.doi.org/10.1787/888933205617

As seen in Figure 5, energy used in transport is dominated by oil products, which account for 94% of all energy used in transport across the 41 countries. Oil represents between 67% of all transport energy in Russia, which uses a high proportion of natural gas in transport, and 100% in Estonia, Israel, Japan, Mexico and South Africa. Diesel and gasoline are the two most commonly used fuels in the transport category, accounting together for 86% of all transport energy and of CO_2 emissions from transport energy. They are also the most dominant fuels in transport in every country considered – and more than 80% of transport energy in 35 countries – although the respective shares of diesel and gasoline used vary between countries.

Outside oil products, natural gas forms a significant proportion of transport energy in several countries, with the highest shares being seen in Russia (33%), Argentina (20%) and the Slovak Republic (18%). Brazil and the Slovak Republic also use a small proportion of biofuels in transport (8% and 7%, respectively). Only Argentina, Brazil, Korea, Russia, the Slovak Republic and Turkey derive more than 20% of their transport energy from non-oil sources.

Figure 9 summarises the composition of transport energy and emissions from transport energy, ordered from those countries with the highest share of diesel and gasoline to total transport energy, to those countries with the least.

3.2.2. *Effective tax rates on transport energy and carbon emissions from energy use in transport*

As seen above, the transport category is taxed more heavily than other categories. This is true both across all countries and within all countries (except for Brazil, where transport energy is untaxed), as seen in the graphical profiles in Part II of this report and in Part II of OECD (2013b).

Although effective tax rates on energy use and carbon emissions from transport energy are higher than on other uses of energy, there is still a wide degree of variation between countries, from zero taxation in Brazil (during the temporary suspension of the CIDE between June 2012 and February 2015) and low effective tax rates in Indonesia and Russia (less than EUR 0.01 per GJ and less than EUR 0.1 per tonne of carbon emissions from energy) to EUR 18.9 per GJ and EUR 263 per tonne of CO_2 in the United Kingdom. Among the other selected partner economies, Argentina and South Africa tax transport energy at rates similar to Poland and Spain and at higher rates than the Americas and Australasia.

As effective tax rates on transport energy are typically higher for oil products than on other fuels, countries with higher shares of non-oil energy in transport have comparatively lower cross-category rates, all else being equal. This is the case, for example, in Turkey, which has the highest observed tax rate on gasoline used for transport purposes, but has a comparatively high share of LPG in transport.

Effective tax rates on transport fuels also vary considerably within countries, both by fuel use and by fuel type. Tables 5 and 6 present average rates on transport energy and on carbon emissions from transport energy, respectively. Information on the contribution of different fuels to the total size of the respective tax bases is also provided. Some fuels used in small amounts are not presented, although they are included in the calculation of the overall rate for transport fuels. Individual country results are set out in Annex B.

Figure 9. **Composition of transport energy and of CO_2 emissions by fuel**

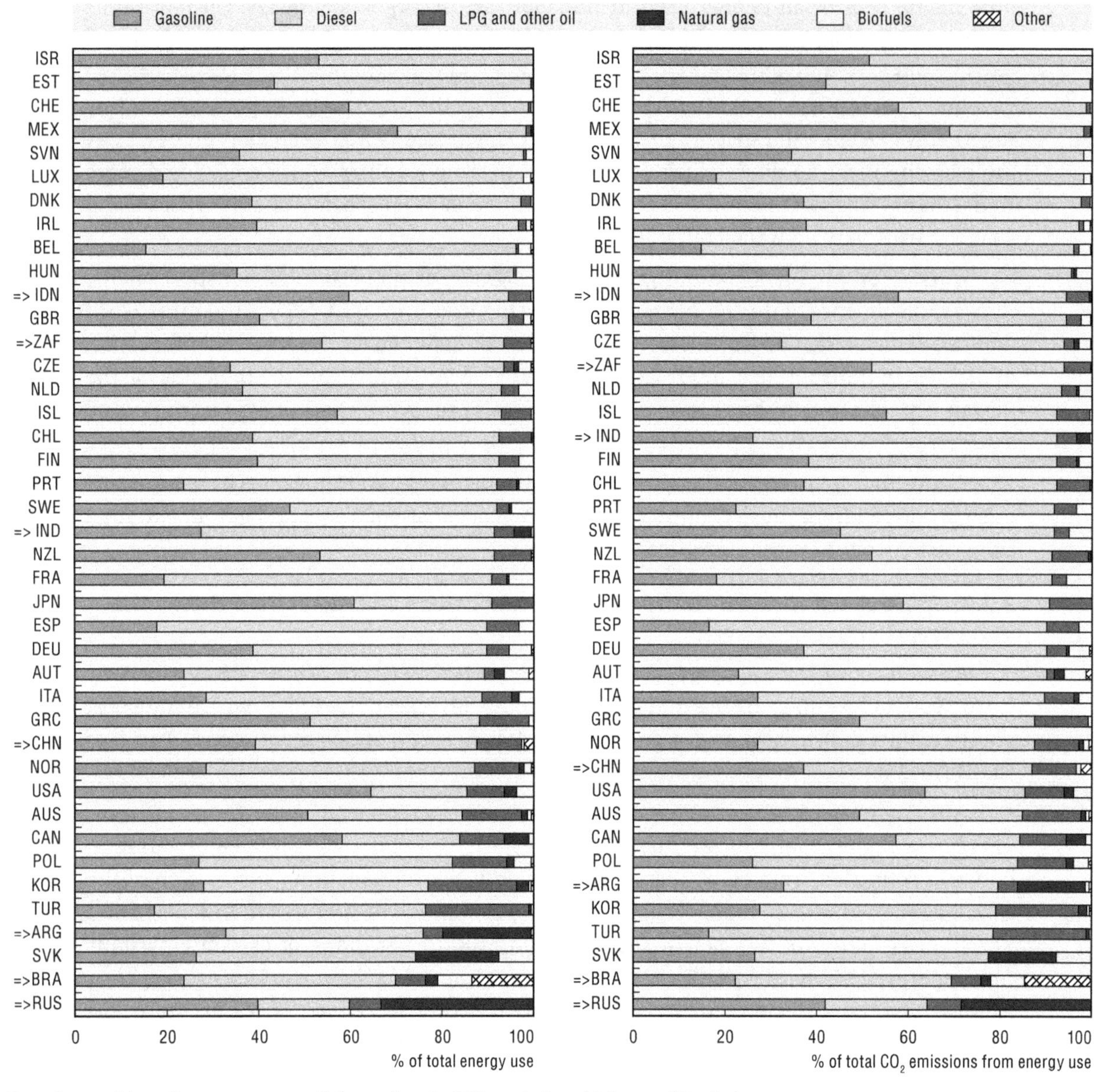

Note: Composition of transport energy (left panel) and of CO_2 emissions (right panel) by fuel.

Source: OECD calculations based on energy use data for 2009 from IEA (2014), *IEA World Energy Statistics and Balances* (database), Doi: http://dx.doi.org/10.1787/data-00513-en.

StatLink http://dx.doi.org/10.1787/888933205622

Across and within all 41 countries considered, except Brazil, road transport is taxed at higher rates than non-road transport, whether considered in terms of energy content or in carbon emissions from transport energy use. Across all 41 countries considered, road energy is taxed on average at EUR 5.6 per GJ (78.8 per tonne of CO_2) as opposed to EUR 0.8 per GJ for non-road fuels (10.8 per GJ). This may be for several reasons. Firstly, oil products are taxed at higher rates – and other sources of energy are more commonly used in non-road transport. However, a more likely reason is that governments deliberately choose to apply higher taxes to road fuels, either for revenue purposes or to address other externalities associated with their use.

Figure 10. **Average effective tax rates on transport energy and on CO_2**

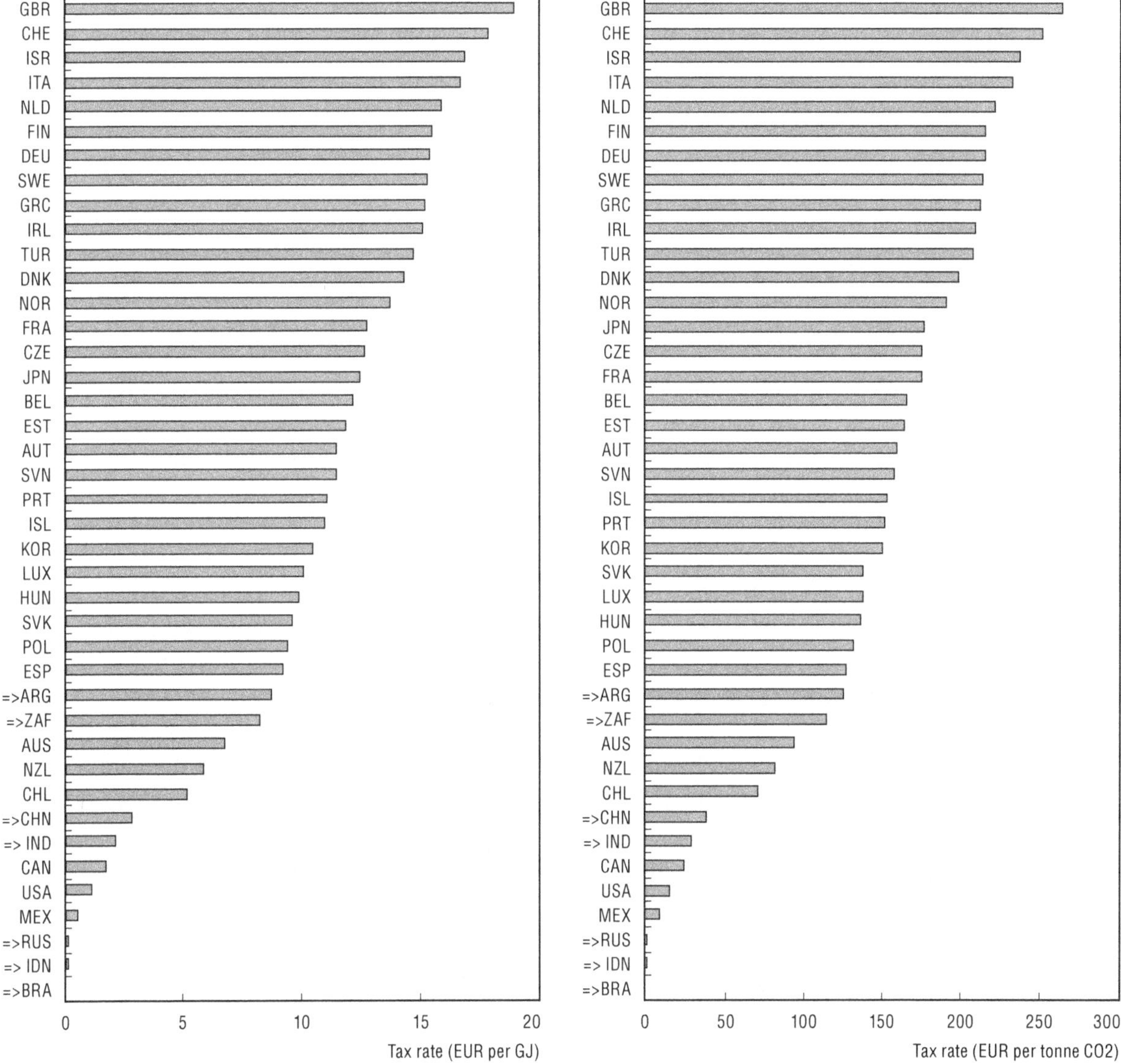

Note: Average effective tax rates on transport energy (left panel) and on CO_2 (right panel).

Source: OECD calculations for selected partner economies; OECD (2013b), *Taxing Energy Use – A Graphical Analysis*, Doi: http://dx.doi.org/10.1787/9789264183933-en, for all other countries. Tax rates are as of 1 April 2012 (except 1 July 2012 for AUS and BRA and 4 April 2012 for ZAF); energy use data is for 2009 from IEA (2014), *IEA World Energy Statistics and Balances* (database), Doi: http://dx.doi.org/10.1787/data-00513-en. Figures for CAN, IND and USA include only federal taxes.

StatLink http://dx.doi.org/10.1787/888933205631

Table 5. **Weighted average effective tax rates on transport energy by fuel type and use (EUR per GJ)**

		Gasoline	Diesel	LPG	Aviation fuels	Biofuels	Natural gas	All fuels
	% of base	49%	37%	1%	6%	3%	3%	100%
Road use	87	5.18	6.47	5.86	0.00	4.47	0.54	5.60
Non road use	13	1.20	1.81	2.93	0.40	0.00	0.04	0.75
Total transport use	100	5.16	6.06	5.84	0.40	4.46	0.12	4.96

Source: OECD calculations for selected partner economies; OECD (2013b), *Taxing Energy Use – A Graphical Analysis*, Doi: http://dx.doi.org/10.1787/9789264183933-en, for all other countries. Tax rates are as of 1 April 2012 (except 1 July 2012 for AUS and BRA and 4 April 2012 for ZAF); energy use data is for 2009 from IEA (2014), *IEA World Energy Statistics and Balances* (database), Doi: http://dx.doi.org/10.1787/data-00513-en. Figures for CAN, IND and USA include only federal taxes.

StatLink http://dx.doi.org/10.1787/888933205961

Table 6. **Weighted average effective tax rates on CO_2 from transport energy by fuel type and use (EUR per tonne CO_2)**

		Gasoline	Diesel	LPG	Aviation fuels	Biofuels	Natural gas	All fuels
	% of base	48%	38%	1%	6%	3%	3%	100%
Road use	87	74.69	87.33	92.81	0.00	63.08	9.69	78.76
Non road use	13	17.34	24.48	46.51	5.64	0.00	0.67	10.83
Total transport use	100	74.41	81.84	92.57	5.64	63.07	2.13	70.05

Source: OECD calculations for selected partner economies; OECD (2013b), *Taxing Energy Use – A Graphical Analysis*, Doi: http://dx.doi.org/10.1787/9789264183933-en, for all other countries. Tax rates are as of 1 April 2012 (except 1 July 2012 for AUS and BRA and 4 April 2012 for ZAF); energy use data is for 2009 from IEA (2014), *IEA World Energy Statistics and Balances* (database), Doi: http://dx.doi.org/10.1787/data-00513-en. Figures for CAN, IND and USA include only federal taxes.

StatLink http://dx.doi.org/10.1787/888933205978

Among the fuels used in the transport category, gasoline and diesel are the highest taxed in energy terms, both across and within all countries (excluding Brazil). The lower relative carbon content of LPG means that in carbon terms, the LPG used for transport faces the highest tax rate. LPG accounts for only 1% of all energy used in transport, however. By contrast, other transport fuels are taxed at much lower rates (for example, natural gas, which is taxed at EUR 0.1 per GJ and EUR 2.1 per tonne of CO_2) and in many countries are not taxed. With the exception of Argentina, the selected partner economies do not tax either natural gas for transport use or energy used in domestic aviation. On average, biofuels (mostly ethanol and biodiesel) are taxed at around two-thirds of the rates applying to oil products. The underlying treatment is quite diverse, likely reflecting differing views as to the net carbon impact of biofuels and the role of non-tax policies like blending requirements as well as other policy objectives including industry support or energy supply concerns. The result is that a few countries tax biofuels at rates equivalent to the energy product they replace, some exempt them from taxation entirely, and many tax them at concessionary rates.

While a unit of gasoline is taxed at a lower rate than a unit of diesel (EUR 5.2 per GJ against EUR 6.5 per GJ, or EUR 74.7 per tonne of CO_2 against EUR 87.3 per tonne of CO_2), the opposite is the case when countries are considered individually, where diesel is taxed at a lower rate than gasoline in both energy and carbon terms in all countries except Brazil (where transport fuels were *de facto* untaxed due to the temporary suspension of the CIDE tax between June 2012 and February 2015) and the United States, where diesel is taxed at slightly higher rates. In most countries, the lower effective tax rates on diesel in energy and carbon terms are due to tax rates per litre for both fuels which are lower for diesel. The different characteristics of both fuels – diesel fuel has roughly 10% more energy and 18% more carbon emissions per litre than gasoline – mean that these differences are greater when effective tax rates are measured in energy terms and greater still if measured in carbon terms.

Between countries, the simple average difference between tax rates on gasoline and diesel is 32% in energy terms and 37% in carbon terms. Among selected partner economies, the difference is greatest in India, where diesel is taxed at an effective tax rate that is 66% lower than gasoline in energy terms and 68% lower in carbon terms, and smallest in Indonesia, where diesel is taxed at an effective tax rate that is 4% lower in energy terms and 10% lower in carbon terms, although effective tax rates on both fuels in Indonesia are very low. Among the selected partner economies, Argentina is the only country to report a tax expenditure in respect of the lower tax rate on diesel. Figure 11 shows the tax rates on gasoline and diesel in all countries on an energy basis.

Figure 11. **Effective tax rates on gasoline and diesel for road use**

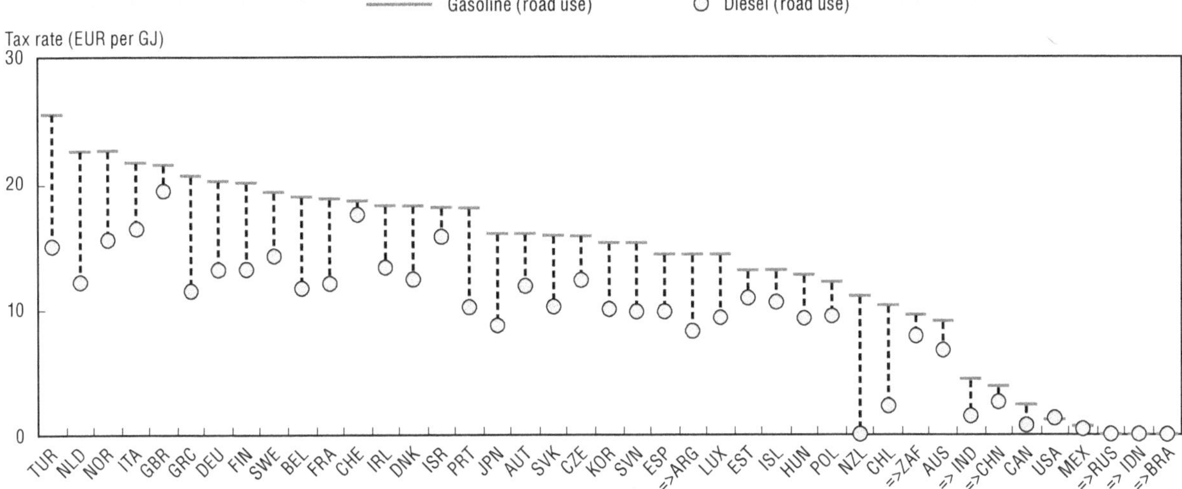

Source: OECD calculations for selected partner economies; OECD (2013b), *Taxing Energy Use – A Graphical Analysis*, Doi: http://dx.doi.org/10.1787/9789264183933-en, for all other countries. Tax rates are as of 1 April 2012 (except 1 July 2012 for AUS and BRA and 4 April 2012 for ZAF); energy use data is for 2009 from IEA (2014), *IEA World Energy Statistics and Balances* (database), Doi: http://dx.doi.org/10.1787/data-00513-en. Figures for CAN, IND and USA include only federal taxes. NZL applies a road-user charge to diesel vehicles on a per-kilometre basis which is not included in the figure.

StatLink http://dx.doi.org/10.1787/888933205644

From an environmental perspective, the lower tax rates for diesel relative to gasoline are not warranted, even where diesel vehicles are more fuel efficient than their gasoline counterparts, due to the higher levels of air pollutants, CO_2 and other social costs (e.g. accidents, congestion and noise) associated with a litre of diesel (Harding, 2014).

The lower effective tax rates on diesel in 39 countries are considered alongside the respective use of diesel and gasoline in each country. Figure 12 shows on the horizontal axis the size of the road diesel tax base in each country in terms of energy, relative to that of gasoline. A number greater than 100% indicates that a country uses more diesel than gasoline. Similarly, on the vertical axis, the graph shows the effective tax rate on diesel in energy terms as a percentage of the effective tax rate on gasoline, with a number above 100% representing a higher tax rate on diesel than gasoline. With the exception of the United States (in the upper left hand corner) and Brazil (at 0 on the horizontal axis), as noted above, all countries apply a higher effective tax rate to diesel than gasoline for road use. Twenty-eight countries are shown in the lower right-hand quadrant, where there is both a lower effective tax rate on diesel and a higher share of carbon emissions from diesel than gasoline, with the difference being the most marked in Belgium, France, Luxembourg and Spain. Among the selected partner economies, Argentina, Brazil and India have higher levels of carbon emissions from diesel than from gasoline; while the differential in diesel taxation relative to gasoline taxation is highest among selected partner economies in Argentina, India and Russia.

3.3. *Taxation of heating and process use of energy*

3.3.1. *Users and sources of heating and process energy*

The heating and process category includes energy used for industrial production and energy transformation as well as energy used for commercial and residential heating. Across all 41 countries considered, heating and process energy accounts for around 40%

of energy use and around 50% of carbon emissions from energy use; ranging from 20% of energy use (and 14% of emissions from energy use) in Israel to 56% of energy use in Indonesia and 71% of carbon emissions from energy use in the Slovak Republic. Across all energy considered, 64% of heating and process energy and emissions are accounted for by industrial production or energy transformation, with the remainder being accounted for by the residential and commercial sectors.

Figure 12. **Use and taxation of diesel for road use relative to gasoline (GJ)**

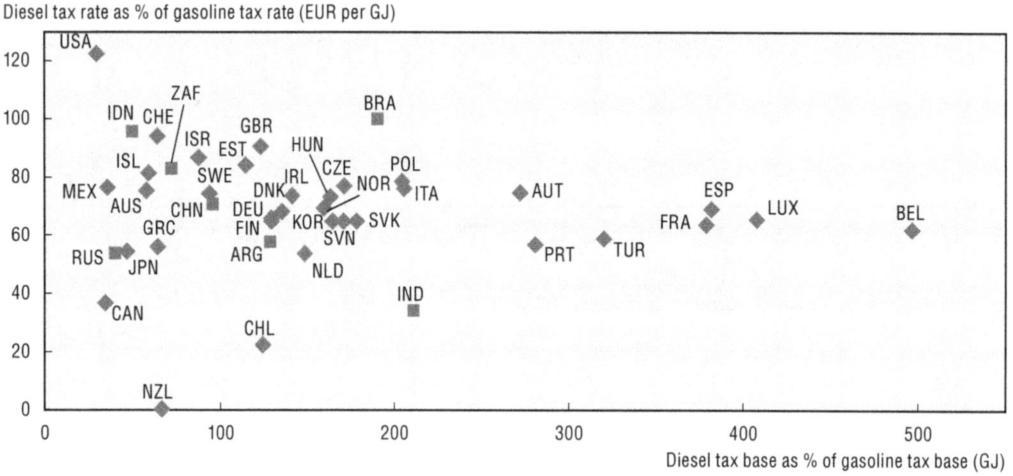

Diesel tax rate as % of gasoline tax rate (EUR per GJ)

Diesel tax base as % of gasoline tax base (GJ)

Source: OECD calculations for selected partner economies; OECD (2013b), *Taxing Energy Use – A Graphical Analysis*, Doi: http://dx.doi.org/10.1787/9789264183933-en, for all other countries. Tax rates are as of 1 April 2012 (except 1 July 2012 for AUS and BRA and 4 April 2012 for ZAF); energy use data is for 2009 from IEA (2014), *IEA World Energy Statistics and Balances* (database), Doi: http://dx.doi.org/10.1787/data-00513-en. Figures for CAN, IND and USA include only federal taxes. NZL applies a road-user charge to diesel vehicles on a per-kilometre basis which is not included in the figure. BRA is shown at 100% as taxes on both gasoline and diesel were suspended between June 2012 and February 2015.

StatLink http://dx.doi.org/10.1787/888933205655

As shown in Figure 5, heating and process energy is derived from a more diverse range of fuels than transport energy. The most common forms of heating and process energy are coal and natural gas, at 30% of all heating and process energy each, although the higher carbon content of coal means that coal is responsible for 38% of carbon emissions from heating and process fuel use compared to 20% from natural gas. Coal is intensively used by South Africa, China and Poland, where it accounts for 50% or more of heating and process fuel, with a lesser share in the other countries considered. Twenty-two of the forty-one countries source less than 10% of their heating and process energy from coal, including Argentina, Brazil and Indonesia.

Other fuels used in the generation of heating and process energy include oil products (22% of all heating and process energy and 18% of emissions, of which the largest share is from diesel, at 8% of heating and process energy and 7% of emissions), although the share of oil products varies considerably between countries from 8% in the Czech Republic and South Africa to 67% in Israel. The remainder of heating and process energy is derived from biomass – most significantly in Brazil, India, Indonesia and Sweden – and renewables, although these account for only a very small share of heating and process energy in total, at 0.5% of all energy use. Higher shares are seen in Iceland (70%) and Israel (26%); in no other country does the proportion of renewable energy to total heating and process energy exceed 5%.

Fossil fuels therefore account for 82% of all energy use in the heating and process category across the 41 countries considered. They are the source of more than 75% of energy use in this category in all countries except Austria, Brazil, Denmark, Estonia, Finland, Iceland, Indonesia, Israel, Portugal, New Zealand, Sweden and Switzerland.

Figure 13 summarises the composition of energy for heating and process use and emissions from energy for heating and process use, ordered from those countries with the highest share of coal to total heating and process energy, to those countries with the least.

Figure 13. **Composition of heating and process energy and of CO_2 emissions from heating and process energy**

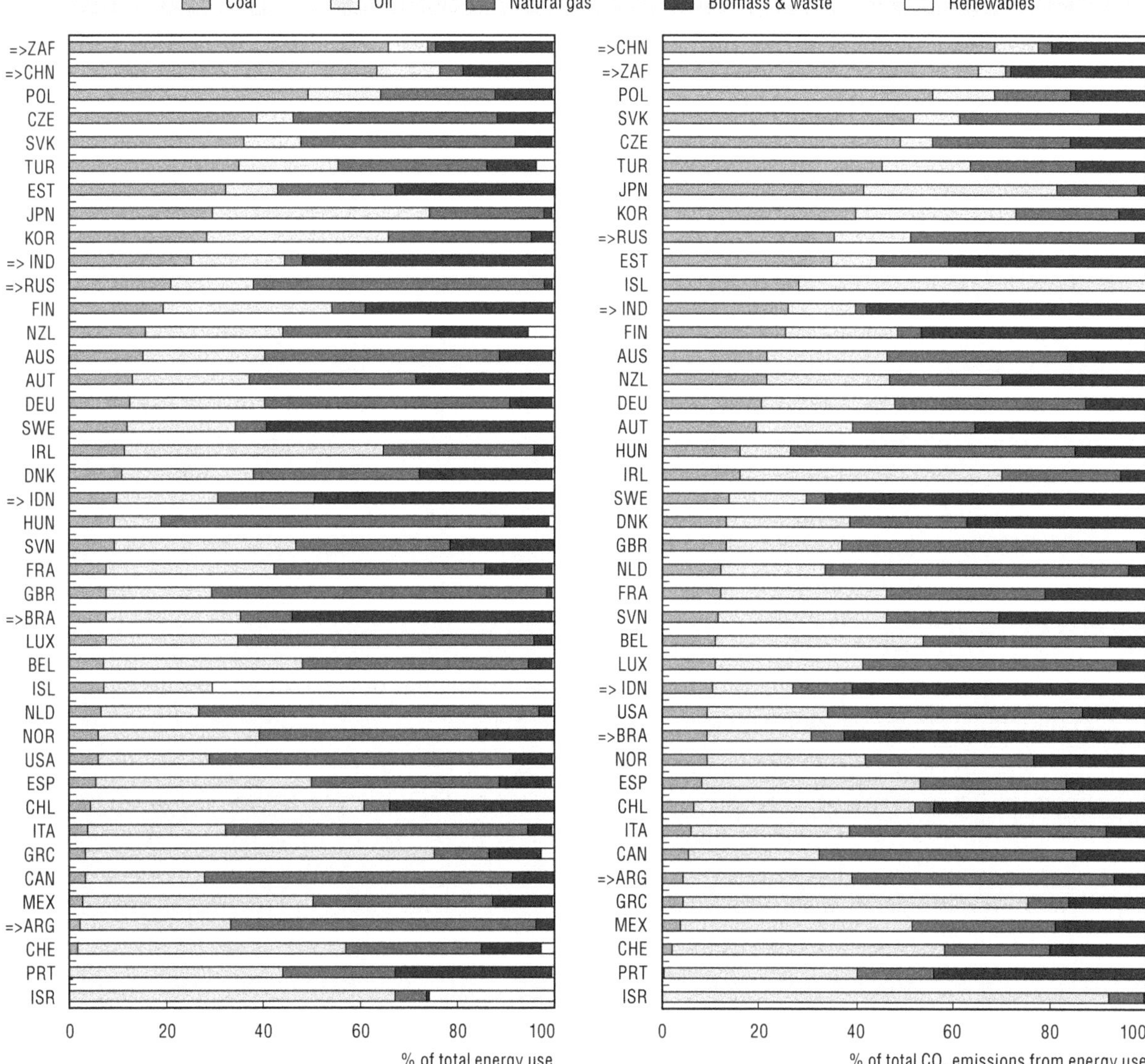

Note: Composition of heating and process energy (left panel) and of CO_2 emissions from heating and process energy (right panel).

Source: OECD calculations based on energy use data for 2009 from IEA (2014), *IEA World Energy Statistics and Balances* (database), Doi: http://dx.doi.org/10.1787/data-00513-en.

StatLink http://dx.doi.org/10.1787/888933205667

3.3.2. Effective tax rates on heating and process energy and carbon emissions from energy for heating and process use

Drawing on the graphical profiles for the 41 countries (presented in Part II of this report and in Part II of OECD, 2013b), heating and process energy is shown to be taxed at lower rates, and less consistently, than energy used in transport. There is wide variation between countries in their approach to the taxation of energy in this category, both in terms of the rates applied and in the sources and uses of energy which are taxed. As a result, effective tax rates in this category also vary widely, from being slightly negative (effectively a subsidy of EUR 0.01 per GJ and EUR 0.10 per tonne CO_2) in Chile, as a result of a petroleum price stabilisation scheme, to EUR 2.61 per GJ in Ireland and EUR 42.25 per tonne of CO_2 in Israel. Among the selected partner economies, rates range from EUR 0 per GJ and tonne of CO_2 in Brazil and Indonesia (and less than EUR 0.001 per GJ and EUR 0.01 per tonne of CO_2 in Russia), who together with the United States do not tax energy used in heating and process use, to EUR 0.2 per GJ and EUR 3 per GJ in Argentina. EU member countries, which are subject to the EU's Energy Tax Directive, apply nineteen of the twenty highest tax rates in this category, even before the price signal provided by the EU ETS (which has not been incorporated into these effective tax rates, as described in Section 2) is taken into account.

Most of the American and Asian countries tax heating and process fuel more lightly (Argentina, Australia, China, India, Japan, Mexico and New Zealand) or do not tax heating and process energy (Brazil, Indonesia and the United States).

Within the selected partner economies there is a wide variation in the tax rates applied to fuels and in the fuels that are taxed. Argentina, India, China and South Africa tax some oil products for heating and process use, at rates ranging from around EUR 1.45 per GJ in India to EUR 72.28 per GJ in South Africa. Argentina also taxes natural gas for heating and process purposes, while India taxes coal used in this category at a low rate. Tables 7 and 8 summarise, on an energy and carbon basis respectively, the effective tax rates on all energy used in this category, broken down by broad categories of fuel use and fuel type. Fuel use is divided into residential and commercial use on the one hand and industrial and energy transformation use (e.g. oil refineries) on the other. Information on the shares of different fuels in the respective tax bases is also provided. Some fuels used in small amounts are not presented in a separate column though they are included in the overall rate for all fuels. Individual country results are again provided in Annex B.

Across all energy use considered, diesel has the highest tax rate in energy and carbon terms, being taxed, on average, at EUR 1.6 per GJ and EUR 22.2 per tonne of CO_2. Other oil products are also taxed at higher rates. Natural gas is taxed at much lower rates, on average, at EUR 0.2 per GJ and EUR 3.8 per tonne of CO_2. The lowest tax rate on heating and process energy is seen for coal, at EUR 0.04 per GJ and EUR 0.5 per tonne of carbon emissions from energy.

The average unit of energy for industrial and energy transformation purposes is taxed lower than for residential and commercial use, particularly in carbon terms, which may result from the different profile of fuels used in each category; with lower taxed fuels, particularly coal, being more common in industrial and energy transformation use. The distinction between tax rates on industrial and energy transformation use of fuels and fuels used for commercial and residential heating varies by country. This pattern holds for the different fuels shown in Tables 7 and 8, with the exception of coal and fuel oil for residential use, which are taxed at lower rates than the same fuel for industrial and energy transformation purposes.

Figure 14. **Average effective tax rates on heating and process energy and on CO_2**

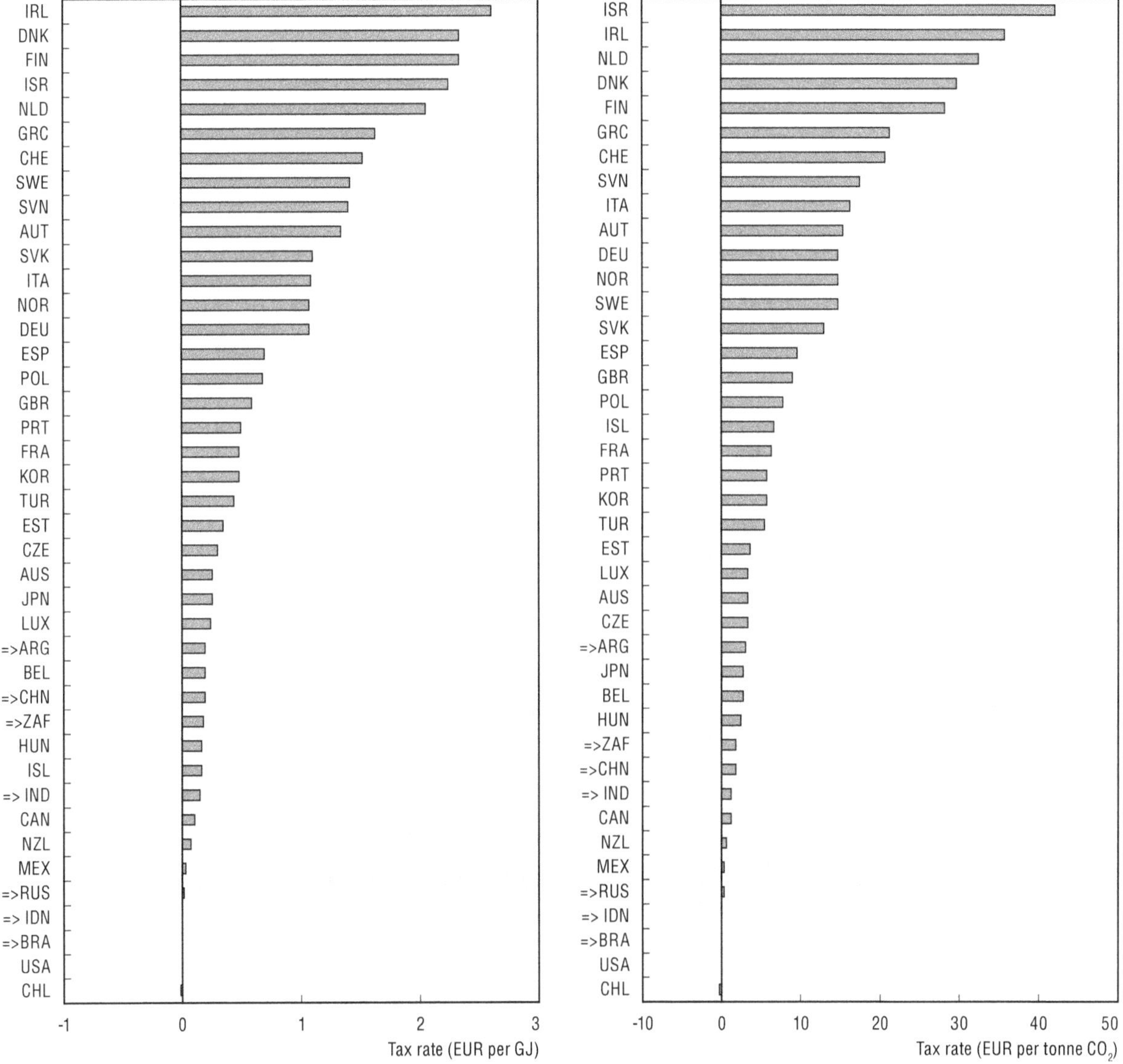

Note: Average effective tax rates on heating and process energy (left panel) and on CO_2 (right panel).

Source: OECD calculations for selected partner economies; OECD (2013b), *Taxing Energy Use – A Graphical Analysis*, Doi: http://dx.doi.org/10.1787/9789264183933-en, for all other countries. Tax rates are as of 1 April 2012 (except 1 July 2012 for AUS and BRA and 4 April 2012 for ZAF); energy use data is for 2009 from IEA (2014), *IEA World Energy Statistics and Balances* (database), Doi: http://dx.doi.org/10.1787/data-00513-en. Figures for CAN, IND and USA include only federal taxes. The price signals sent by the EU ETS (EU member countries, Iceland and Norway) to some forms of energy use in this category are not included in the figures, but were relatively modest over the time period considered (on average EUR 13 per tonne of CO_2 in 2010-11).

StatLink http://dx.doi.org/10.1787/888933205675

However, on a country by country basis, the pattern is more varied. Residential and commercial use of fuels is taxed at higher rates in energy terms than industrial fuels in 21 countries (22 in carbon terms); with industrial use being taxed more highly in 17 countries in energy terms (16 in carbon terms). The difference is most pronounced in Israel, Denmark and the Netherlands, where residential rates are higher than industrial rates and in Sweden and Ireland, where industrial rates are higher than residential rates. In several other countries the difference is very small (Chile, Estonia, Greece, Mexico, Russia, the Slovak Republic and Spain). Figures 15 and 16 show effective tax rates on energy for both uses, in energy and carbon terms respectively.

Table 7. **Weighted average effective tax rates on heating and process energy by fuel type and use (EUR per GJ)**

		Coal	Natural gas	Diesel	Fuel oil	Other oil products	All fuels
	% of base	30%	30%	8%	2%	11%	100%
Residential and commercial use	36	0.04	0.30	1.79	0.67	0.42	0.33
Industrial and energy transformation use	64	0.05	0.14	1.54	0.97	0.12	0.22
Total heating and process use	100	0.05	0.21	1.64	0.95	0.21	0.26

Source: OECD calculations for selected partner economies; OECD (2013b), *Taxing Energy Use – A Graphical Analysis*, Doi: http://dx.doi.org/10.1787/9789264183933-en, for all other countries. Tax rates are as of 1 April 2012 (except 1 July 2012 for AUS and BRA and 4 April 2012 for ZAF); energy use data is for 2009 from IEA (2014), *IEA World Energy Statistics and Balances* (database), Doi: http://dx.doi.org/10.1787/data-00513-en. Figures for CAN, IND and USA include only federal taxes. The price signals sent by the EU ETS (EU member countries, Iceland and Norway) to some forms of energy use in this category are not included in the figures, but were relatively modest over the time period considered (on average EUR 13 per tonne of CO_2 in 2010-11).

StatLink http://dx.doi.org/10.1787/888933205988

Table 8. **Weighted average effective tax rates on CO_2 from heating and process energy by fuel type and use (EUR per tonne CO_2)**

		Coal	Natural gas	Diesel	Fuel oil	Other oil products	All fuels
	% of base	38%	20%	7%	2%	9%	100%
Residential and commercial use	34	0.42	5.27	24.21	8.70	6.38	4.01
Industrial and energy transformation use	66	0.48	2.54	20.78	12.58	1.70	2.58
Total heating and process use	100	0.47	3.75	22.19	12.22	3.08	3.07

Source: OECD calculations for selected partner economies; OECD (2013b), *Taxing Energy Use – A Graphical Analysis*, Doi: http://dx.doi.org/10.1787/9789264183933-en, for all other countries. Tax rates are as of 1 April 2012 (except 1 July 2012 for AUS and BRA and 4 April 2012 for ZAF); energy use data is for 2009 from IEA (2014), *IEA World Energy Statistics and Balances* (database), Doi: http://dx.doi.org/10.1787/data-00513-en. Figures for CAN, IND and USA include only federal taxes. The price signals sent by the EU ETS (EU member countries, Iceland and Norway) to some forms of energy use in this category are not included in the figures, but were relatively modest over the time period considered (on average EUR 13 per tonne of CO_2 in 2010-11).

StatLink http://dx.doi.org/10.1787/888933205999

Figure 15. **Effective tax rates on energy: Residential and commercial vs. industrial and energy transformation use**

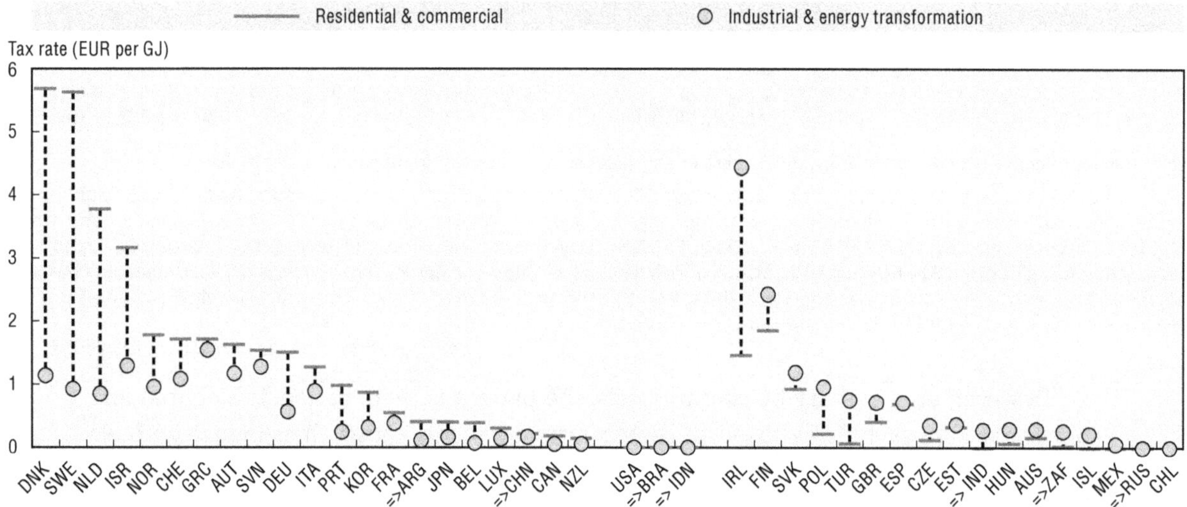

Source: OECD calculations for selected partner economies; OECD (2013b), *Taxing Energy Use – A Graphical Analysis*, Doi: http://dx.doi.org/10.1787/9789264183933-en, for all other countries. Tax rates are as of 1 April 2012 (except 1 July 2012 for AUS and BRA and 4 April 2012 for ZAF); energy use data is for 2009 from IEA (2014), *IEA World Energy Statistics and Balances* (database), Doi: http://dx.doi.org/10.1787/data-00513-en. Figures for CAN, IND and USA include only federal taxes. The price signals sent by the EU ETS (EU member countries, Iceland and Norway) to some forms of energy use in this category are not included in the figures, but were relatively modest over the time period considered (on average EUR 13 per tonne of CO_2 in 2010-11).

StatLink http://dx.doi.org/10.1787/888933205689

Figure 16. **Effective tax rates on CO_2: Residential and commercial vs. industrial and energy transformation use**

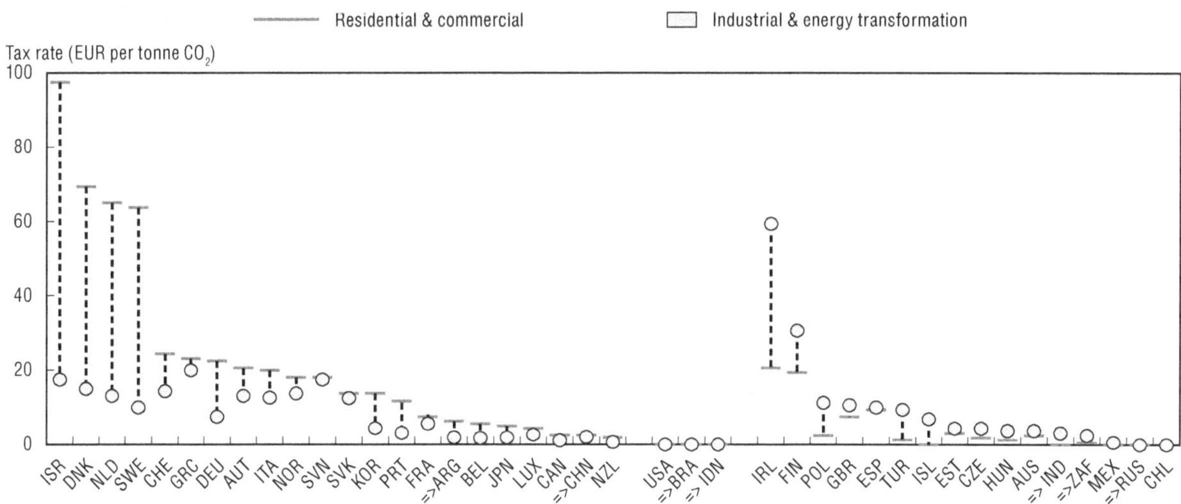

Source: OECD calculations for selected partner economies; OECD (2013b), *Taxing Energy Use – A Graphical Analysis*, Doi: http://dx.doi.org/10.1787/9789264183933-en, for all other countries. Tax rates are as of 1 April 2012 (except 1 July 2012 for AUS and BRA and 4 April 2012 for ZAF); energy use data is for 2009 from IEA (2014), *IEA World Energy Statistics and Balances* (database), Doi: http://dx.doi.org/10.1787/data-00513-en. Figures for CAN, IND and USA include only federal taxes. The price signals sent by the EU ETS (EU member countries, Iceland and Norway) to some forms of energy use in this category are not included in the figures, but were relatively modest over the time period considered (on average EUR 13 per tonne of CO_2 in 2010-11).

StatLink http://dx.doi.org/10.1787/888933205690

The typically lower tax rates in the heating and process category together with the large variation in rates between different users may be the result of deliberate policy choices to lower rates for that particular sector, perhaps prompted by competitiveness or distributional concerns. Another possible explanation for the different rates applied to different users of heating and process energy, which is particularly likely when the difference in the rates of both groups is small, is that the different fuels used by each group could result in this difference. Among the countries considered, coal and oil products are used more heavily in industrial processes than in domestic or commercial use. Conversely, natural gas is the most common residential and commercial heating fuel, together with biomass. The different tax rates that apply to different fuels may therefore result in different effective tax rates for different users of these fuels.

The heating and process category contains the most diverse use of fossil fuels of all three use categories. Given the different ways and frequencies with which the different fuels are taxed in this category, it is interesting to compare the taxation of the three main sources of fossil fuels (oil, coal and natural gas) within this category.

Figure 17 sets out along the horizontal axis the cumulative percentage of energy across all 41 countries considered derived from each of the three sources of fossil fuels, ranked from the lowest to the highest taxed. The vertical axis shows the tax rates in EUR per GJ that applies to each of the three fuels at each percentage of use. The vertical axis is shown in logarithmic terms in order to better display the detail of the figure. This figure shows that coal is the least taxed of all fossil fuel sources of energy in this category, with over 85% untaxed. Reflecting the presence of large coal users among those countries which apply lower or zero tax rates on coal, less than 0.5% of coal is taxed at a rate higher than EUR 10 per GJ, despite the high environmental and other social costs associated with coal. Natural gas is more frequently taxed and typically taxed at higher rates: 72% of energy derived from natural gas is untaxed and 10% is taxed at a rate higher than EUR 10 per GJ. Oil products are taxed at the highest rates and most frequently, with 54% of energy derived from oil products subject to energy taxes and 30% taxed at a rate higher than EUR 10 per GJ.

Figure 17. **Tax rates on fossil fuels used in heating and process use, by cumulative energy use from each fuel**

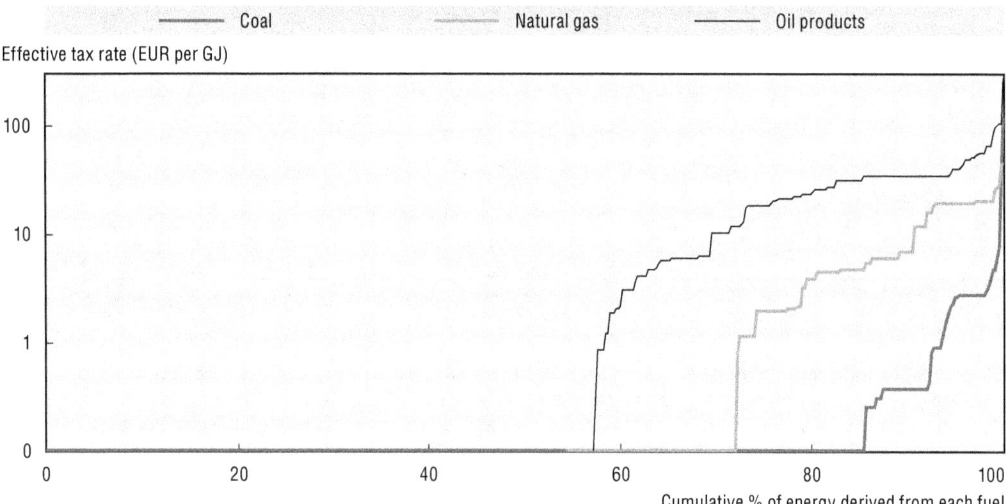

Source: OECD calculations for selected partner economies; OECD (2013b), *Taxing Energy Use – A Graphical Analysis*, Doi: http://dx.doi.org/10.1787/9789264183933-en, for all other countries. Tax rates are as of 1 April 2012 (except 1 July 2012 for AUS and BRA and 4 April 2012 for ZAF); energy use data is for 2009 from IEA (2014), *IEA World Energy Statistics and Balances* (database), Doi: http://dx.doi.org/10.1787/data-00513-en. Figures for CAN, IND and USA include only federal taxes. The price signals sent by the EU ETS (EU member countries, Iceland and Norway) to some forms of energy use in this category are not included in the figures, but were relatively modest over the time period considered (on average EUR 13 per tonne of CO_2 in 2010-11). Tax rates are shown on a logarithmic scale on the vertical axis.

StatLink http://dx.doi.org/10.1787/888933205703

This figure does not include, however, the coverage of emission trading schemes, such as the EU and NZ ETS, or regional level ETS schemes such as those that apply in China. Emissions trading schemes are more likely to cover coal and natural gas use than oil products, which would change the picture shown in this graph. However, recent prices under ETS schemes have been comparatively low, relative to energy taxes (World Bank, 2014a).

3.4. *Taxation of energy used to generate electricity*

3.4.1. *Users and sources of energy used to generate electricity*

Energy used to generate electricity is the most diverse in terms of the fuel-mix used across countries. Coal, oil, natural gas, combustibles and renewables are all used in significant proportions in many countries. The most common source of energy used in electricity generation across the countries considered is coal, followed by renewables. The electricity category is the only category of energy use where renewables form a significant part of the energy mix.

Figure 18 shows the sources of energy used to generate electricity in each country, in energy terms. This graph shows the proportion of each energy source used to generate energy (the input energy), rather than the resulting electricity (the output energy) derived from each source. Differences in the efficiency of generation from different fuel sources will mean that fuels which are comparatively more efficient (often renewable forms of electricity generation) will represent a smaller share of the input energy than of the output energy. The converse is true for fuels which are less efficient. The graph is ordered by those countries with the highest share of fossil fuels in electricity generation to those with the lowest share. This ranges from Israel, where 100% of energy used to generate electricity is derived from fossil fuel sources, to Iceland, where over 99% of energy used to generate electricity is from renewable sources.

Figure 18. **Composition of energy used to generate electricity, by input fuel**[1]

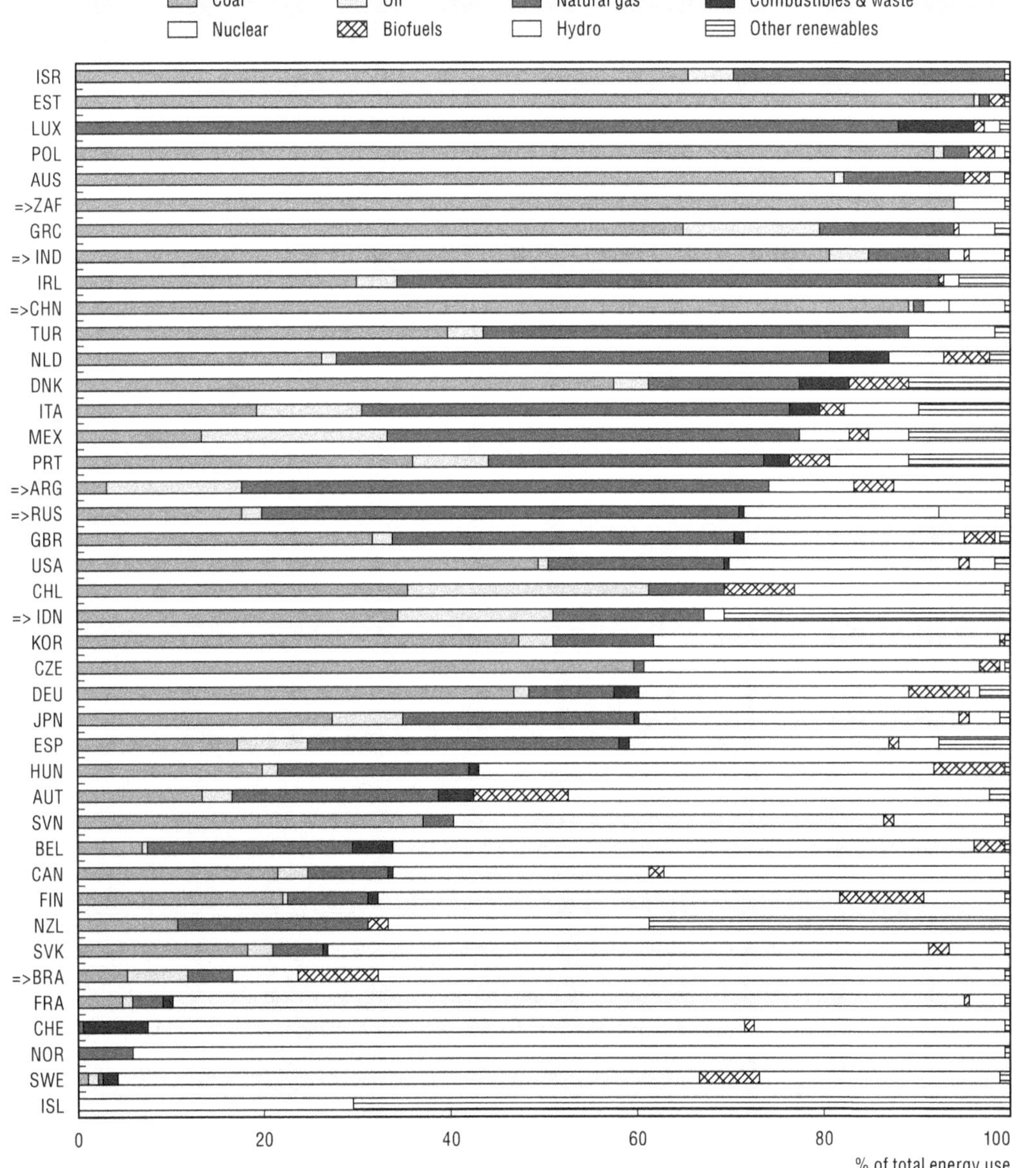

This graph shows the proportion of fuels used as inputs to electricity generation in each country. Electricity is a secondary energy product which is derived from these primary sources. Different fuels and electricity plants have differing ratios of generation efficiency, which means the proportions of electricity derived from each fuel will differ from that fuel's proportion of the input energy.

Source: OECD calculations based on energy use data for 2009 from IEA (2014), *IEA World Energy Statistics and Balances* (database), Doi: http://dx.doi.org/10.1787/data-00513-en.

StatLink http://dx.doi.org/10.1787/888933205718

Coal forms a large proportion of energy used to generate electricity in several countries. It represents more than 50% (and in several cases, more than 80%) of energy used to generate electricity in Australia, China, Denmark, Estonia (oil shale), Greece, India, Israel, Poland and South Africa. Natural gas is the source of over 50% of energy used in electricity generation in Argentina, Ireland, Luxembourg and Russia. Relative to both the transport and heating and process categories, oil accounts for only a small proportion of most countries' energy use in electricity generation. Energy from non-carbon sources, including renewable and nuclear sources forms the majority of energy used in electricity generation in 14 countries (Austria, Belgium, Brazil, Canada, Finland, France, Hungary, Iceland, New Zealand, Norway, the Slovak Republic, Slovenia, Sweden and Switzerland).

The size and share of renewables as inputs to electricity generation differs between countries. Figure 19 shows the total amount of renewable energy used to generate electricity in each country (measured in TJs, on the left-hand axis) against the proportion of renewable energy to total energy used to generate electricity (measured as a % of all energy used to generate electricity), on the right-hand axis.

Figure 19. **Amount and proportion of renewables (excluding nuclear) in electricity generation**

Biofuels | Hydro | Other Renewables | Renewables as share of electricity generation energy (RHS)

Energy (million TJ) | % of total energy used to generate electricity

ISL, NOR, =>BRA, NZL, AUT, CAN, SWE, =>IDN, CHL, CHE, PRT, ITA, FIN, DNK, MEX, =>ARG, SVN, ESP, DEU, TUR, SVK, HUN, IRL, =>RUS, NLD, =>CHN, GRC, JPN, USA, FRA, GBR, AUS, =>IND, POL, LUX, BEL, CZE, EST, KOR, =>ZAF, ISR

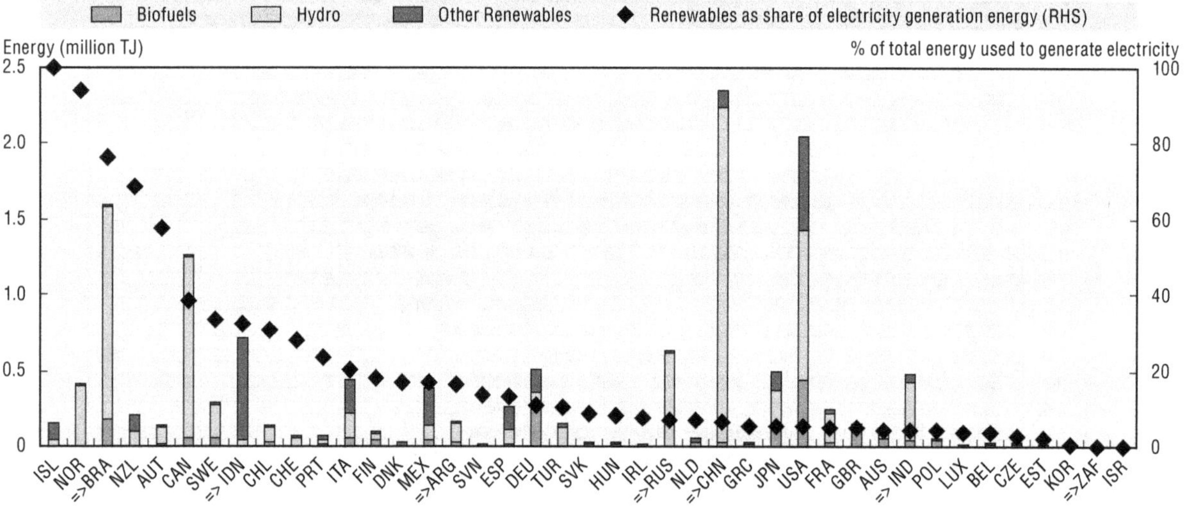

Source: OECD calculations based on energy use data for 2009 from IEA (2014), *IEA World Energy Statistics and Balances* (database), Doi: http://dx.doi.org/10.1787/data-00513-en.

StatLink http://dx.doi.org/10.1787/888933205725

China uses the highest amount of renewables in electricity generation, at 2.5 million TJs (just under 7% of energy used in electricity generation), with the majority of this coming from hydro generation. Significant levels of energy from renewable sources are also found, in decreasing order, in the United States, Brazil, Canada, Indonesia, Russia, India and Germany. When considered as a proportion of energy used to generate electricity, the percentage of renewable energy sources is highest in Iceland, Norway and Brazil (again in decreasing order). The selected partner economies account for 43% of the total amount of renewables used in electricity generation across all 41 countries considered.

In addition, 24 countries generate electricity from nuclear sources. In 2009, the largest user of nuclear energy for electricity generation in absolute terms was the United States (26% of electricity generation fuels). Around half as much nuclear energy is used to generate electricity in France and in Japan. In percentage terms, France uses the highest amount of nuclear energy as a share of all input energy for electricity generation, at 85%. Other countries that use nuclear energy for more than 50% of their input energy for electricity generation, in decreasing order, are the Slovak Republic, Switzerland, Belgium, Sweden and Finland. Among the selected partner economies, India, China, South Africa, Brazil and Argentina use a small proportion of nuclear energy as an input to electricity generation (less than 10%) and Russia uses nuclear energy for around 20% of its inputs to electricity generation.

3.4.2. *Effective tax rates on energy used to generate electricity*

Taxes on electricity may be levied either directly on the fuels used to generate electricity or on the consumption of the resulting electricity. The approach to the taxation of electricity across the 41 countries is mixed, although the taxation of electricity on a consumption basis is more common. This is also true for the selected partner economies, where Argentina, Brazil and India tax electricity consumption. India and South Africa also directly tax coal and some other input fuels used to generate electricity.

As discussed in Section 2.2, the methodology used in this report "looks through" taxes on electricity consumption to estimate the implicit tax rates on the primary energy source used to generate electricity. This approach means that comparatively efficient and low-carbon fuels will have a lower implicit tax rate in energy and carbon terms than those that are less efficient or more carbon-intensive. Where countries tax the primary energy used to generate electricity directly, the tax rate for each energy source is calculated. Where a country taxes both the fuels used to generate electricity and electricity consumption, both levels of taxation are taken into account in calculating the effective tax rate on each primary energy source. In the EU countries, as well as Iceland and Norway, the EU ETS will provide an additional price on carbon emissions from electricity generation from some sources. As described in Section 2, the price signals provided by the EU ETS have not been included in the analysis.

Although taxing electricity consumption is viewed for the purposes of this report as an indirect tax on the fuels used to generate electricity, an electricity tax that does not distinguish between sources of electricity generation does not send any price signal with regard to the fuels used to generate electricity or with regard to the efficiency of generation.

As seen in the other categories, effective tax rates on the different fuels used to generate electricity differ significantly between countries in energy terms. Tax rates are highest in two EU member countries, Denmark and the Netherlands, at more than EUR 6 per GJ of energy used in electricity generation. Several countries (Canada, Chile, Indonesia, New Zealand, Russia and the United States) do not tax either electricity or the fuels used to generate it, and several others (Australia, China and Mexico) have an effective tax rate on energy used to generate electricity that is less than EUR 0.1 per GJ. Of the selected partner economies, Brazil taxes energy used in electricity generation at the highest rates, at EUR 1.56 per GJ. This is the 7th highest rate across all 41 countries considered. Argentina and South Africa tax energy used in electricity generation at EUR 0.33 and EUR 0.24 per GJ, respectively, around the average country rate for the group of countries considered.

Figure 20. **Average effective tax rates on energy used in electricity generation**

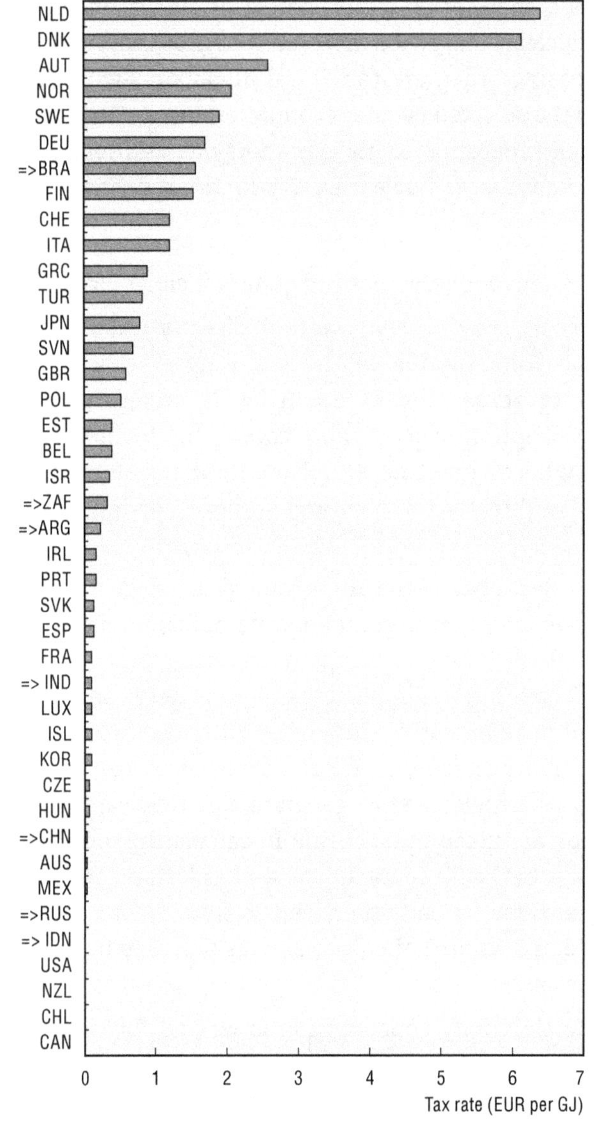

Source: OECD calculations for selected partner economies; OECD (2013b), *Taxing Energy Use – A Graphical Analysis*, Doi: http://dx.doi.org/10.1787/9789264183933-en, for all other countries. Tax rates are as of 1 April 2012 (except 1 July 2012 for AUS and BRA and 4 April 2012 for ZAF); energy use data is for 2009 from IEA (2014), *IEA World Energy Statistics and Balances* (database), Doi: http://dx.doi.org/10.1787/data-00513-en. Figures for CAN, IND and USA include only federal taxes. The price signals sent by the EU ETS (EU member countries, Iceland and Norway) to some forms of energy use in this category are not included in the figures, but were relatively modest over the time period considered (on average EUR 13 per tonne of CO_2 in 2010-11).

StatLink http://dx.doi.org/10.1787/888933205739

As in the other categories of energy use shown in the graphical profiles, different fuels used in electricity generation are taxed at different rates. Tables 9 and 10 show the rates applying to different forms of energy products used to generate electricity, in energy and carbon terms respectively. These take into account both direct taxes on the fuels used to generate electricity and taxes on the consumption of electricity. Due to the more complicated construction of the effective tax rates in this category, they should be interpreted carefully. These rates demonstrate the implicit effective tax rate on the energy (or alternately, carbon content) in electricity if the electricity tax were assumed to be a tax on the input fuels. Consequently, if carbon-intensive fuels form a small proportion of the generation mix, the effective tax rate on carbon thus calculated will be very high. A tax on electricity consumption that does not distinguish between electricity from carbon sources and electricity from non-carbon sources cannot send an effective price signal about the use of carbon-intensive generation sources. Nonetheless, in this report, in order to maintain the same tax coverage for energy and carbon statistics, undifferentiated taxes on electricity consumption are included in the computation of effective tax rates on carbon emissions.

Table 9. **Weighted average effective tax rates on energy used in electricity generation by fuel type (EUR per GJ)**

	Coal	Biofuels	Waste	Natural gas	Oil	Renewables	Hydro	Nuclear	All fuels
% of base	53	1	1	16	3	2	6	18	100
Electricity	0.13	0.64	0.66	0.43	0.50	0.58	0.65	0.27	0.27

Source: OECD calculations for selected partner economies; OECD (2013b), *Taxing Energy Use – A Graphical Analysis*, Doi: http://dx.doi.org/10.1787/9789264183933-en, for all other countries. Tax rates are as of 1 April 2012 (except 1 July 2012 for AUS and BRA and 4 April 2012 for ZAF); energy use data is for 2009 from IEA (2014), *IEA World Energy Statistics and Balances* (database), Doi: http://dx.doi.org/10.1787/data-00513-en. Figures for CAN, IND and USA include only federal taxes. The price signals sent by the EU ETS (EU member countries, Iceland and Norway) to some forms of energy use in this category are not included in the figures, but were relatively modest over the time period considered (on average EUR 13 per tonne of CO_2 in 2010-11).

StatLink http://dx.doi.org/10.1787/888933206005

Table 10. **Weighted average effective tax rates on CO_2 from energy used in electricity generation by fuel type (EUR per tonne CO_2)**

	Coal	Biofuels	Waste	Natural gas	Oil	All fuels
% of base (excl. ISL, NOR, SWE)	80	2	1	14	3	100
Electricity (excl. ISL, NOR, SWE)	2.22	10.77	14.11	5.37	6.67	3.10

Source: OECD calculations for selected partner economies; OECD (2013b), *Taxing Energy Use – A Graphical Analysis*, Doi: http://dx.doi.org/10.1787/9789264183933-en, for all other countries. Tax rates are as of 1 April 2012 (except 1 July 2012 for AUS and BRA and 4 April 2012 for ZAF); energy use data is for 2009 from IEA (2014), *IEA World Energy Statistics and Balances* (database), Doi: http://dx.doi.org/10.1787/data-00513-en. Figures for CAN, IND and USA include only federal taxes. Tax rates and energy use in electricity generation are not included for Iceland, Norway and Sweden. The price signals sent by the EU ETS (EU member countries, Iceland and Norway) to some forms of energy use in this category are not included in the figures, but were relatively modest over the time period considered (on average EUR 13 per tonne of CO_2 in 2010-11).

StatLink http://dx.doi.org/10.1787/888933206019

3.5. Economy-wide effective tax rates

3.5.1. Overview of economy-wide effective tax rates

The heterogeneous patterns of energy use and taxation in each of these categories across the 41 countries considered result in significant differences in the overall level of energy taxation across the 41 countries considered. Figure 21 sets out for each country the overall effective tax rate, on a weighted basis, on energy use (left panel) and on CO_2 emissions from energy use (right panel). Please note that for countries that impose energy taxes at both the federal and provincial level (notably Canada, India and the United States), these figures only account for taxes imposed at the federal level. This is the case for all the results presented in this part of the report.

Figure 21. **Economy-wide average effective tax rates on energy and on CO_2 from energy**

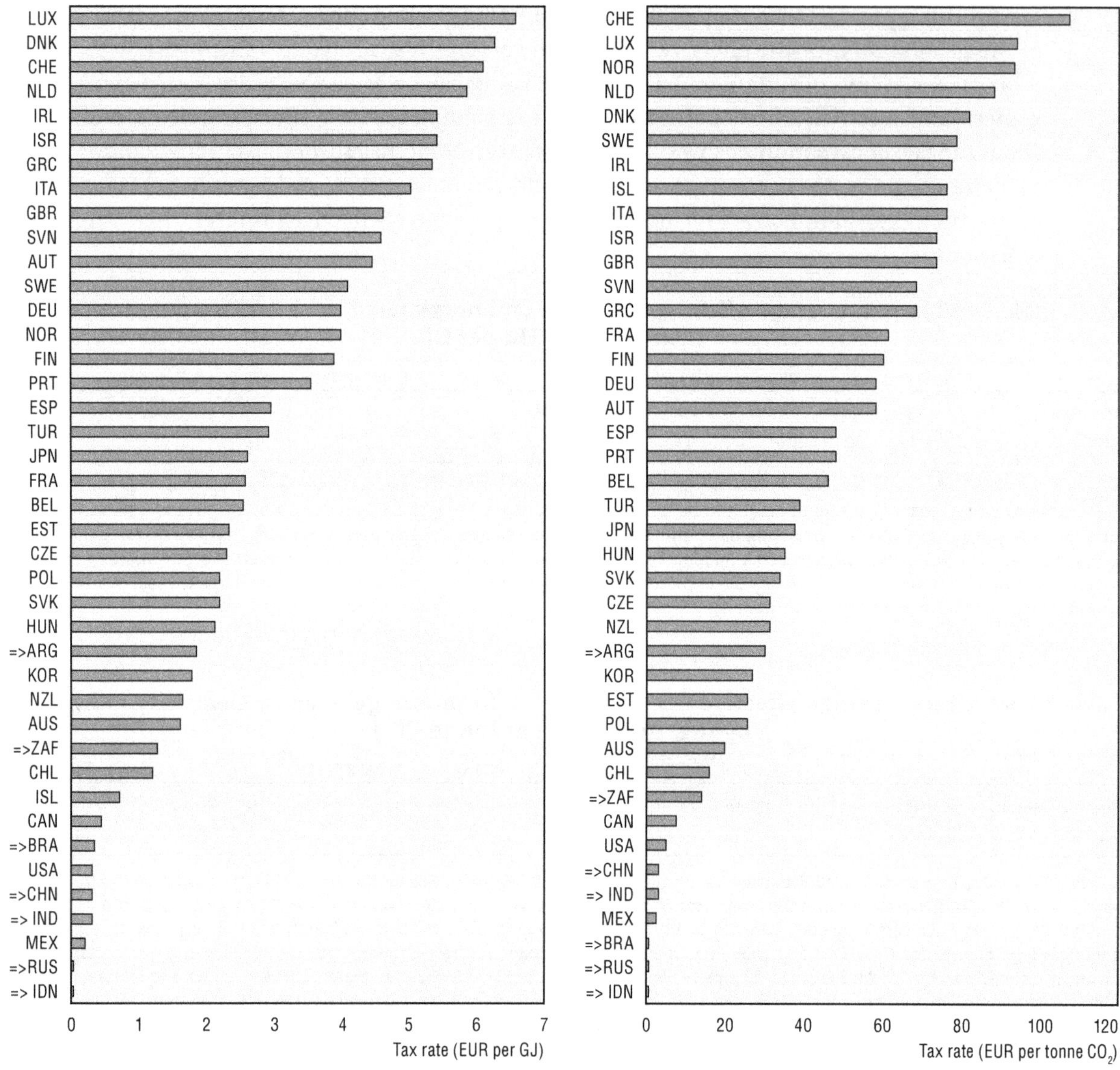

Note: Economy-wide average effective tax rates on energy (left panel) and on CO_2 from energy (right panel).

Source: OECD calculations for selected partner economies; OECD (2013b), *Taxing Energy Use – A Graphical Analysis*, Doi: http://dx.doi.org/10.1787/9789264183933-en, for all other countries. Tax rates are as of 1 April 2012 (except 1 July 2012 for AUS and BRA and 4 April 2012 for ZAF); energy use data is for 2009 from IEA (2014), *IEA World Energy Statistics and Balances* (database), Doi: http://dx.doi.org/10.1787/data-00513-en. Figures for CAN, IND and USA include only federal taxes.

StatLink http://dx.doi.org/10.1787/888933205748

In energy terms, the variation in country averages is very wide, from effective tax rates of less than EUR 0.001 per GJ in Indonesia and Russia, to EUR 6.58 per GJ in Luxembourg. Luxembourg has the highest rate even though its tax rates on most fuel products are not among the highest. Although transport tax rates in Luxembourg are low compared to transport tax rates in neighbouring jurisdictions, the resulting high proportion of transport to total energy generates a high overall tax rate. This is because transport tax rates in Luxembourg are higher than the rates applied to heating and process and electricity in other jurisdictions. The variation in rates among the selected partner economies is narrower: in addition to Indonesia and Russia, economy-wide tax rates on energy in Brazil,

China and India range from EUR 0.28 and EUR 0.35 per GJ and in South Africa and Argentina are EUR 1.25 and EUR 1.84 per GJ, respectively.

Similarly, there is a wide range of effective tax rates on CO_2, when measured on an economy-wide basis, as set out in the right panel of Figure 21. Consistent with the approach of this report these figures take into account all specific taxes on energy whether or not they are explicitly intended to tax carbon emissions. The lowest economy-wide effective tax rates on CO_2 are found in Indonesia and Russia (EUR 0.002 and EUR 0.006 respectively). China and India have an average effective tax rate on carbon emissions from energy of EUR 3.4 and 3.12 per tonne CO_2, respectively, while South Africa has an effective tax rate of EUR 13.86 per tonne CO_2.

In addition to the underlying tax rates, the size of the different use categories in each country, and particularly the share of transport in total energy use, influences the economy-wide tax rates. Since the transport category is taxed at higher rates in every country except Brazil, the share of transport energy in total energy will influence the economy-wide effective tax rates, as seen for Luxembourg. Conversely, countries with a comparatively small share of transport energy, notably China, Iceland, India, Russia and South Africa, will tend to have slightly lower rates than they would if the shares of transport energy were standardised between countries.

The highest overall tax rates on CO_2 tend to be seen in countries which are members of the European Union. In these countries, energy tax policy is significantly shaped by the 2003 EU Energy Taxation Directive, which sets minimum tax rates for a wide range of energy commodities. Many countries with the highest effective tax rates on CO_2 in Figure 21 are countries with explicit carbon taxes (e.g. Denmark, Iceland, Ireland, Norway, Sweden and Switzerland). Eastern European countries tend to have lower effective tax rates on CO_2 and Australasia, the Americas, and Asian economies have the lowest tax rate and typically tax only a small share of total energy use.

3.5.2. *Economy-wide effective tax rates on fuels*

The economy-wide effective tax rates shown for each country in Section 3.5.1 are the result of the differing tax rates applied to different fuels and users of fuel within each country. As shown above, tax rates on oil products are typically higher than those applied to other energy products. This is particularly the case in transport energy, where oil, being the dominant fuel for road use, is taxed at comparatively high rates relative to other fuels and users. Coal and natural gas, which are primarily used for heating and process and electricity generation energy rather than in transport, are taxed at much lower rates for all uses, and coal in particular is often untaxed. Renewables are almost exclusively used in electricity generation and are subject only to the implicit taxes from consumption taxes on energy.

These patterns can be seen in Figures 22 and 23, which set out the effective tax rate for each major fuel group in each country, together with the economy-wide effective tax rates set out in Figure 21. The size of the circle for each fuel represents the share of that fuel in total energy use (Figure 22) and total carbon emissions (Figure 23). Fuels representing less than 5% of total energy use in that country were excluded for clarity. These graphs illustrate the patterns described above. In all countries except Brazil, oil is taxed at the highest effective rate, and is typically the only tax rate above the economy-wide effective tax rate, demonstrating the role of transport in determining economy-wide averages. Conversely, most other fuels are taxed at comparatively low rates, or at zero.

Figure 22. **Summary of energy use and effective tax rates on energy for each fuel and on an economy-wide basis**

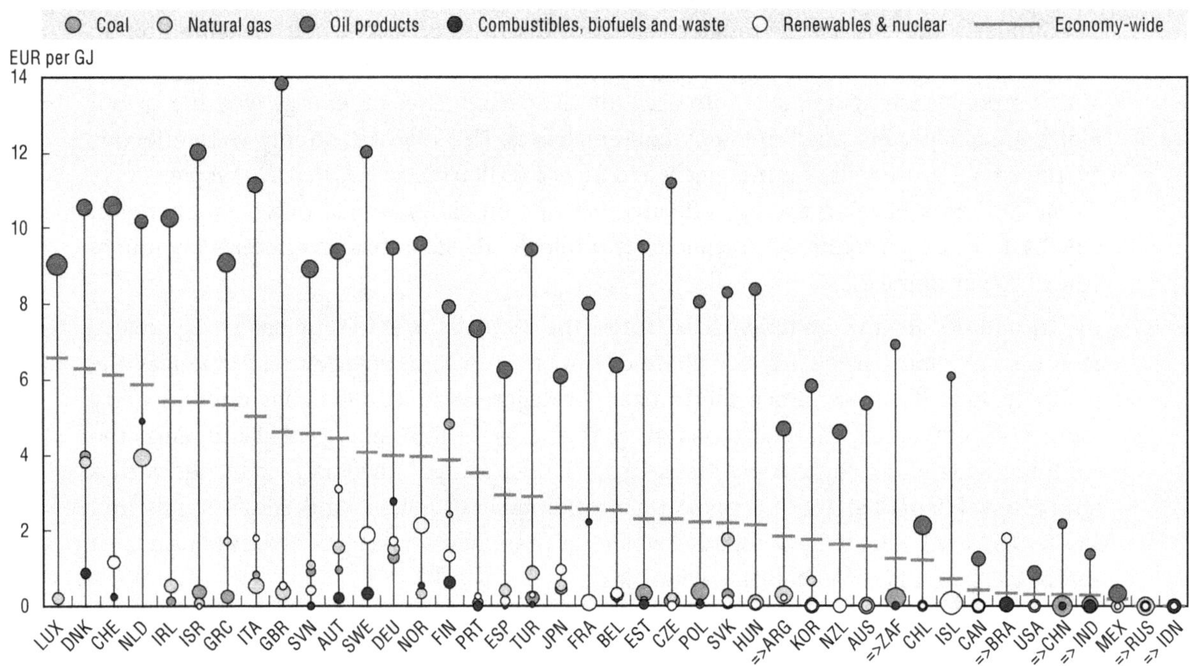

Source: OECD calculations for selected partner economies; OECD (2013b), *Taxing Energy Use – A Graphical Analysis*, Doi: http://dx.doi.org/10.1787/9789264183933-en, for all other countries. Tax rates are as of 1 April 2012 (except 1 July 2012 for AUS and BRA and 4 April 2012 for ZAF); energy use data is for 2009 from IEA (2014), *IEA World Energy Statistics and Balances* (database), Doi: http://dx.doi.org/10.1787/data-00513-en. Figures for CAN, IND and USA include only federal taxes.

StatLink http://dx.doi.org/10.1787/888933205758

3.5.3. *Tax rates on energy and CO_2 – Summary squares profiles*

The full detail of the energy tax picture in each country is set out in the detailed graphical profiles of energy use and taxation shown in Part II of this report, and in *Taxing Energy Use – A Graphical Analysis* (OECD, 2013b). This information can however be condensed into a summary square of energy use and taxation in each country. Figure 24 shows these summary squares of the effective tax rates in terms of energy and carbon content in each of the 41 countries considered. Each of the squares represents the total amount of energy use in each country. The shading shows the proportion of energy subject to each tax rate, from EUR 0 per GJ (white) to over EUR 25 per GJ (black). The dot in each box shows the economy-wide average effective tax rate in that country. Its placement indicates where the mean sits in the percentile distribution of effective tax rates.

As for the summary statistics above, these summary squares highlight the wide variation in effective tax rates on carbon both within and across country economies. Typically, the highest tax rates in each country, the darkest shaded area, are the tax rates on oil products in transport.

3.6. *Energy, GDP and population*

To better understand the different patterns of energy use and taxation observed across the 41 countries considered, these patterns are put into the broader context provided by other economic and demographic indicators, particularly GDP (adjusted for purchasing power parity, PPP) and population.

Figure 23. **Summary of carbon emissions from energy use and effective tax rates on carbon from energy use for each fuel and on an economy-wide basis**

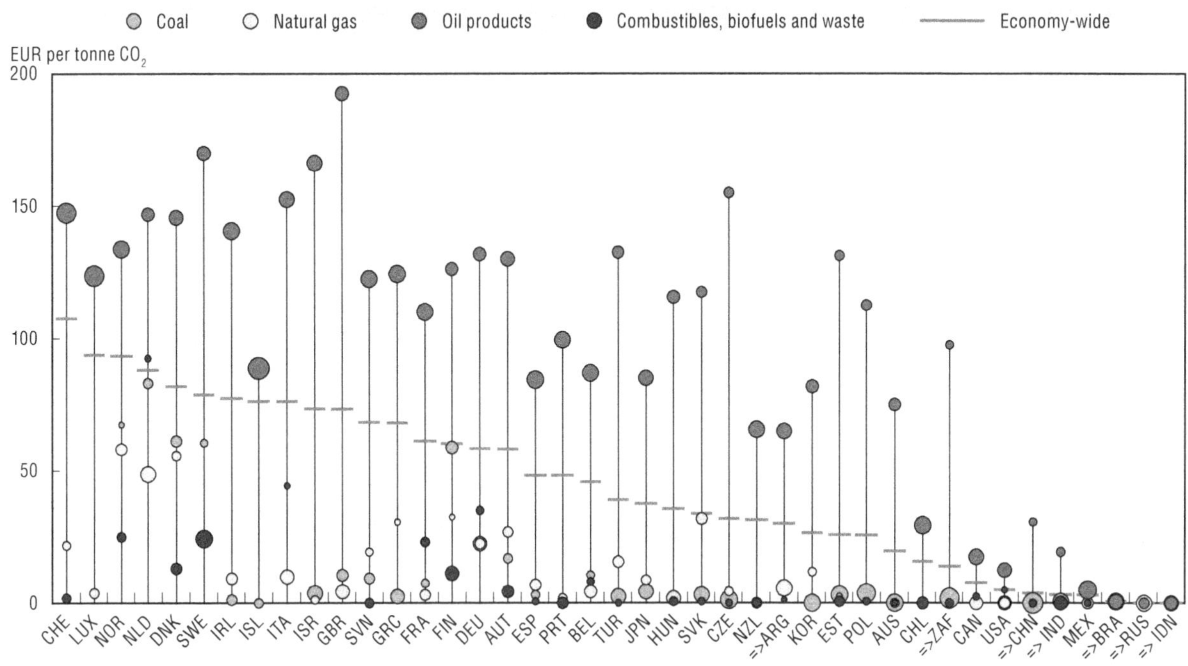

Source: OECD calculations for selected partner economies; OECD (2013b), *Taxing Energy Use – A Graphical Analysis*, Doi: http://dx.doi.org/10.1787/9789264183933-en, for all other countries. Tax rates are as of 1 April 2012 (except 1 July 2012 for AUS and BRA and 4 April 2012 for ZAF); energy use data is for 2009 from IEA (2014), *IEA World Energy Statistics and Balances* (database), Doi: http://dx.doi.org/10.1787/data-00513-en. Figures for CAN, IND and USA include only federal taxes.

StatLink http://dx.doi.org/10.1787/888933205768

The countries considered in this report have vastly different levels of energy use, population and GDP. Comparing the total level of these variables between countries is therefore not helpful in understanding broader patterns of energy use in these countries. However, the relationships between these variables can be compared across countries: energy use or carbon emissions per capita or the energy or carbon efficiency of GDP allow a common basis of consideration across countries. The relationship between these variables differs over time, as described in Section 1.3. The relationships between energy use, carbon emissions from energy use, population and GDP are summarised in Figure 25.

Figure 26 shows, for each of the 41 countries considered, the amount of energy used per capita (on the horizontal axis) against the carbon efficiency of energy use in that country (on the vertical axis). The area under and to the left of each country's position is therefore the level of carbon emissions per capita in each country. A country with a low level of energy use per capita may, for example, have the same level of emissions per capita as a country with higher energy use per capita but a less carbon intensive energy mix: compare for example Denmark and Sweden, which both have around 10 tonnes of carbon emission per person but have very different characteristics of carbon intensity and energy use per capita.

Figure 24. **Summary squares of energy taxation in each country**

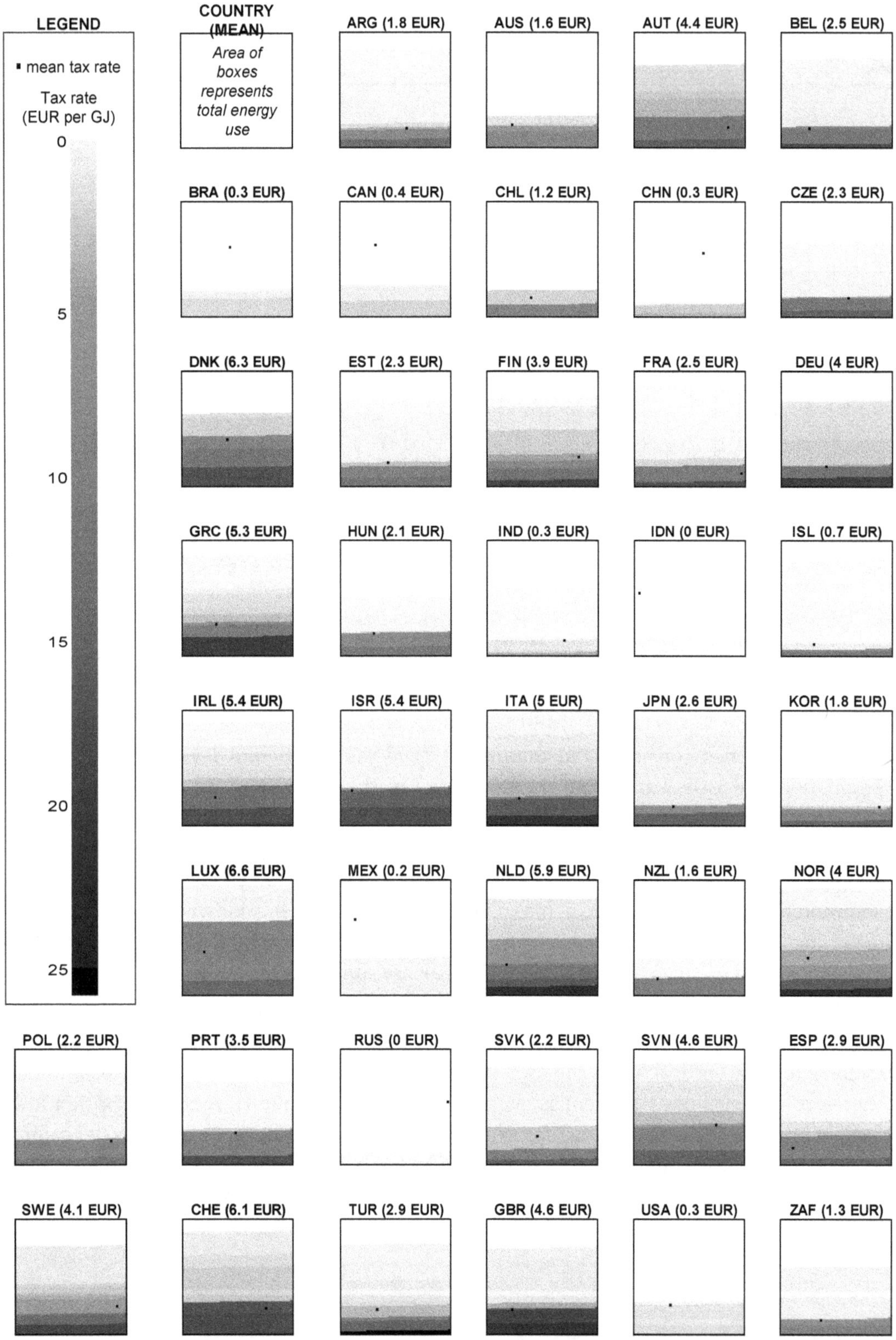

Source: OECD calculations for selected partner economies; OECD (2013b), adapted from *Taxing Energy Use – A Graphical Analysis*, Doi/ http://dx.doi.org/10.1787/9789264183933-en, for all other countries. Tax rates are as of 1 April 2012 (except 1 July 2012 for AUS and BRA and 4 April 2012 for ZAF); energy use data is for 2009 from IEA (2014), *IEA World Energy Statistics and Balances* (database), Doi: http://dx.doi.org/10.1787/data-00513-en. Figures for CAN, IND and USA include only federal taxes.

StatLink http://dx.doi.org/10.1787/888933205775

Figure 25. **Relationships between energy use, carbon emissions from energy use, population and GDP**

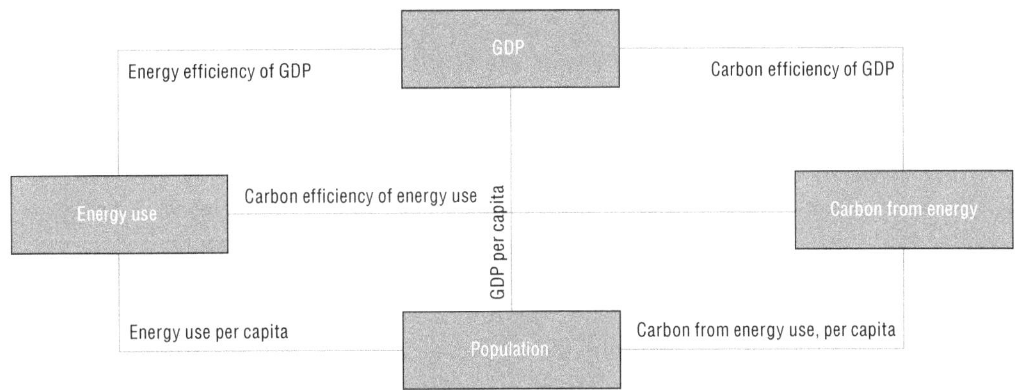

Figure 26. **Energy use per capita and carbon intensity of energy use**

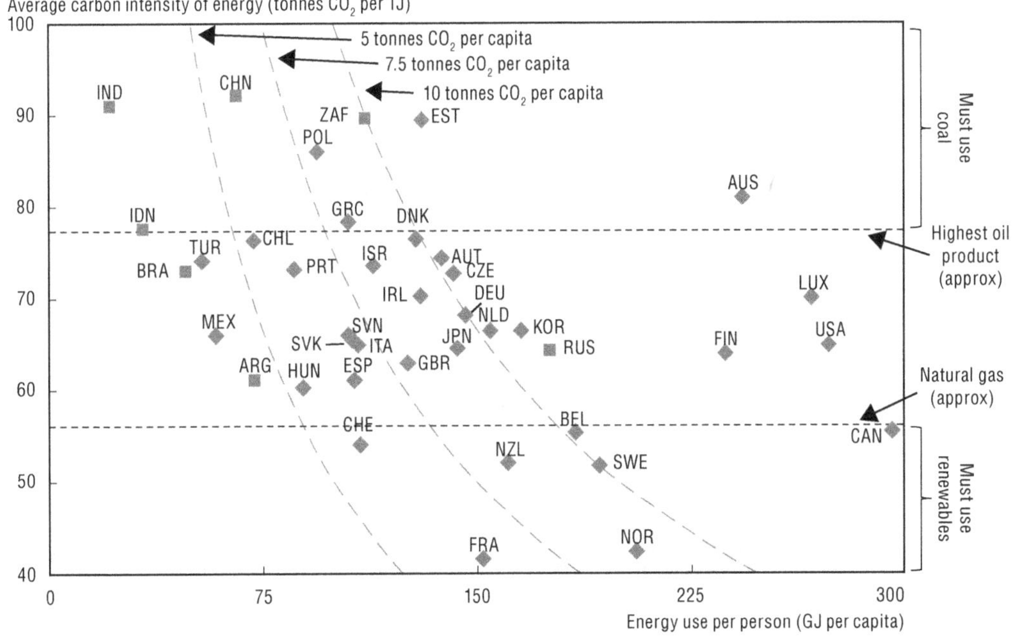

Source: OECD calculations for selected partner economies; OECD (2013b), *Taxing Energy Use – A Graphical Analysis*, Doi: http://dx.doi.org/10.1787/9789264183933-en, and the *World Development Indicators* (World Bank, 2014b).

StatLink http://dx.doi.org/10.1787/888933205784

On the figure, the curved dotted lines show the position on the graph of 5, 7.5 and 10 tonnes of CO_2 per person, respectively, for different levels of carbon intensity and energy use per capita. The horizontal dotted lines show the approximate carbon intensities of natural gas and the most carbon intensive oil product. Natural gas has one of the lowest carbon intensities of any energy source, other than nuclear and renewable energy sources. Therefore, any country which is positioned below this line must have a significant share of renewables or nuclear as part of their energy mix. Similarly, the upper horizontal line denotes the carbon intensity of fuel oil, the highest most-commonly used oil product. Countries above this line therefore must have a significant proportion of coal in their energy mix. A position between these two lines is less clear-cut; countries in this area may have a mix of any energy source, including coal or renewables, but their overall carbon intensity is not dominated by either renewables or coal.

Effective tax rates can also be considered against these broader economic characteristics. Figure 27 shows a simple scatterplot of the effective tax rates on CO_2 emissions from energy use in the 41 countries against their respective carbon intensities of GDP (measured in tonnes of carbon per USD million, adjusted for purchasing power parity). Carbon intensity in relation to GDP is determined by both the relative energy efficiency of GDP and the carbon intensity of the energy mix in each country. Countries with a relatively high carbon intensity of GDP are shown towards the right-hand side, while countries with a relatively low carbon intensity of GDP are located towards the left-hand side of the figure. High carbon intensity of GDP is usually due to relatively low energy-efficiency or due to a relatively carbon-intensive fuel mix (for example, countries that use a high proportion of coal are shown toward the right of the figure), while countries at the left-hand side of the figure are either more energy efficient or have a relatively low-carbon fuel mix (for example, due to a high share of renewable or nuclear energy).

Figure 27. **Average effective tax rates on CO_2 from energy and carbon intensity of GDP (PPP-adjusted)**

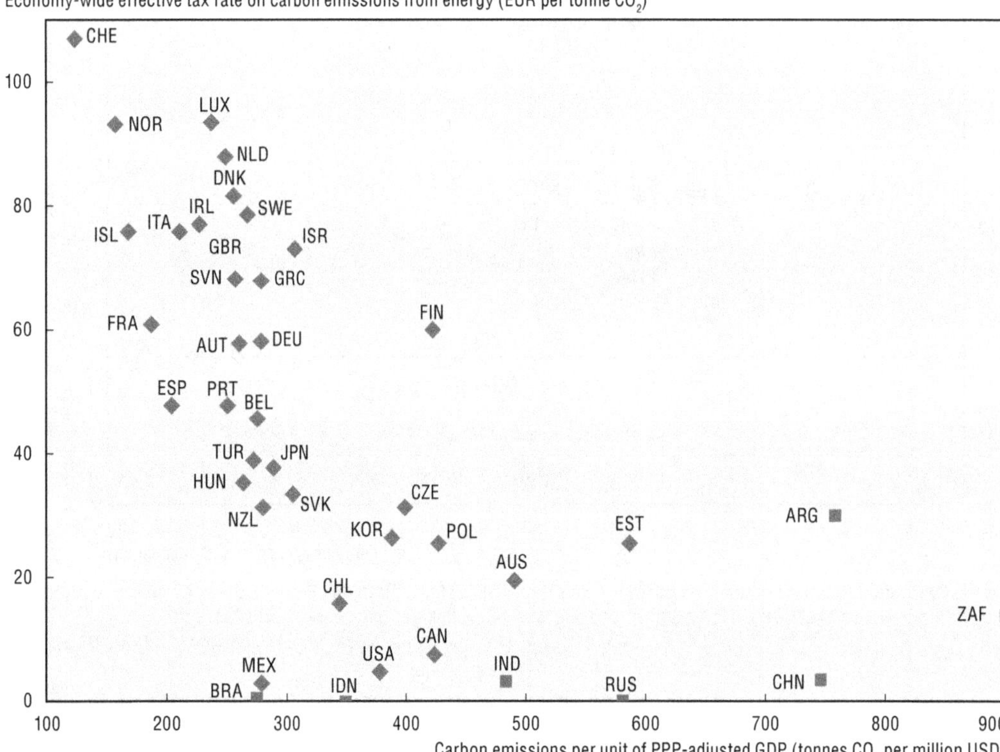

Source: OECD calculations for selected partner economies; adapted from OECD (2013b), *Taxing Energy Use – A Graphical Analysis*, Doi: http://dx.doi.org/10.1787/9789264183933-en, for all other countries. Tax rates are as of 1 April 2012 (except 1 July 2012 for AUS and BRA and 4 April 2012 for ZAF); energy use data is for 2009 from IEA (2014), *IEA World Energy Statistics and Balances* (database), Doi: http://dx.doi.org/10.1787/data-00513-en. Figures for CAN, IND and USA include only federal taxes.

StatLink http://dx.doi.org/10.1787/888933205791

Figure 27 indicates that countries with higher average effective tax rates on CO_2 tend to have less carbon-intensive economies (measured as lower carbon emissions per unit of GDP). While this correlation does not imply causation, it suggests that there is a linkage. It also does not indicate the direction of any causation that may exist. A low carbon intensity may be the result of increased efficiency or different fuel mixes used, partially as a result of higher taxes on carbon. Conversely, a country which has a carbon-intensive GDP may find it more feasible to have a higher tax rate on carbon.

Countries with lower tax rates on carbon less clearly fit the pattern observed among other countries. This may be for two reasons. Firstly, in many of these countries, particularly in the selected partner economies, government intervention in energy markets through non-tax pricing measures means that the energy taxes shown on the right-hand axis of Figure 27 are a less complete indication of the government's energy policy settings than in the other countries. A further possible explanation is that many countries with low tax rates on carbon emissions from energy use have lower per capita incomes, which appear to be positively correlated with energy tax rates, as shown in Figure 28.

Figure 28. **Average effective tax rates on CO_2 from energy and GDP (PPP-adjusted) per capita**

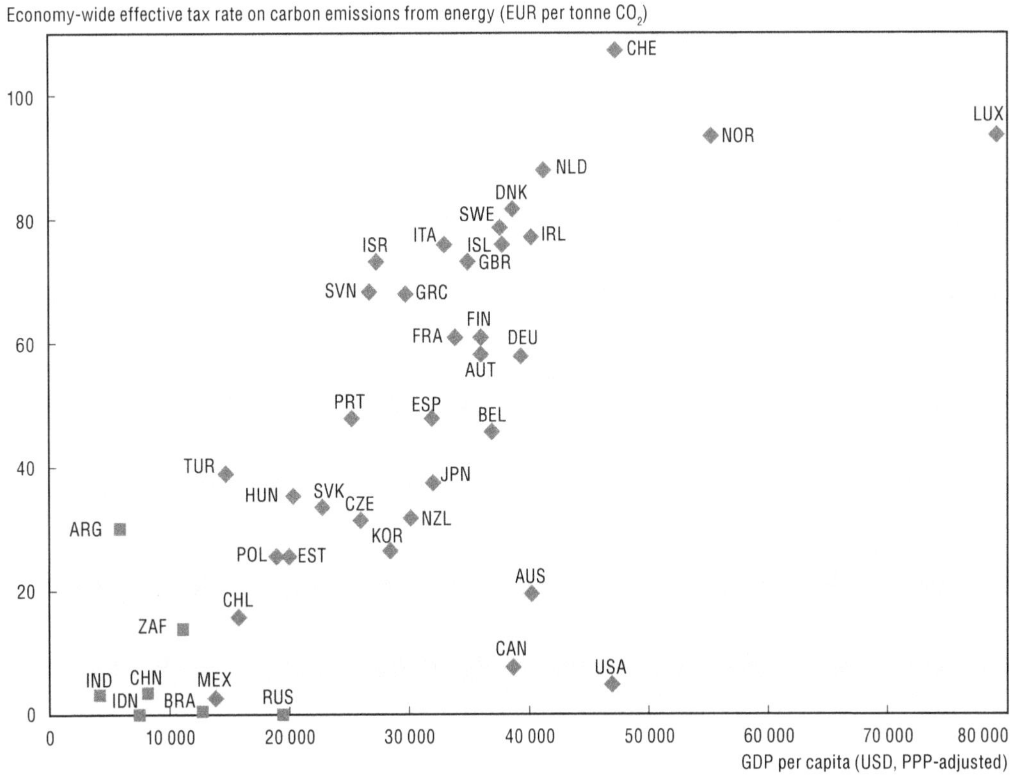

Source: OECD calculations for selected partner economies; OECD (2013b), *Taxing Energy Use – A Graphical Analysis*, Doi: http://dx.doi.org/10.1787/9789264183933-en, for all other countries. Tax rates are as of 1 April 2012 (except 1 July 2012 for AUS and BRA and 4 April 2012 for ZAF); energy use data is for 2009 from IEA (2014), *IEA World Energy Statistics and Balances* (database), Doi: http://dx.doi.org/10.1787/data-00513-en. Figures for CAN, IND and USA include only federal taxes.

3.7. Conclusions

Energy taxes have an important impact on energy use patterns, economic outcomes and the environment through their impact on the overall and relative prices of energy products. The cross-country analysis presented in this report has highlighted the wide diversity of tax rates that apply both across and within the 41 countries for which graphical profiles have been presented in either this report or in *Taxing Energy Use – A Graphical Analysis* (OECD, 2013b). As well as taxes on energy use, many non-tax policies, together with differential rates of value-added taxes, also impact the prices of energy products for different fuels and fuel users. Considering taxes against this broader backdrop is important in understanding the broader policy signals provided in respect of energy use.

The 41 countries analysed in this chapter represent just over 80% of global energy use, and 84% of carbon emissions from energy use, in 2009. The seven selected partner economies that are the focus of this report account for around 36% of energy use in 2009, a share that has grown quickly and is expected to rise rapidly in the coming decades – along with these countries' weight in the world economy.

The selected partner economies discussed in this report are not a homogenous group in terms of their energy use or taxation patterns and should not be construed as such. Nonetheless, certain common themes emerge across several of these countries. On an economy-wide basis, the selected partner economies are among the jurisdictions which have comparatively low average effective taxes on energy use, relative to the full group of 41 countries discussed in this chapter. Underpinning this economy-wide picture, some commonalities emerge in the treatment of different fuels or fuel users. Several of the selected partner economies tax coal at very low rates, or do not tax coal at all, despite its comparatively high environmental and other social costs. Across all 41 countries considered, with the sole exception of Brazil, road transport is taxed at higher rates than other uses of energy in the same jurisdiction, although road transport rates also vary considerably between countries. Diesel for road use is taxed at lower rates in energy terms than gasoline for the same purpose in 39 of the countries considered, including in all selected partner economies except Brazil.

Taxes are one means by which governments can influence the prices of energy products. In addition, as discussed in the introductory chapter, the impact of differential VAT rates and of the main non-tax energy policies influencing producer prices is also important in understanding the broader price signals provided in respect of energy use. Many countries, including some selected partner economies, allow lower VAT rates on some types of energy. The use of non-tax pricing instruments is more common in some of the selected partner economies than it is on average. These policies often, but not always, result in relatively low end-user prices for energy. Considering both the effective tax rates on energy and the non-tax pricing measures, the emerging picture for several of these countries is one of moderate to low end-user prices for energy.

In practice, energy taxation and pricing policies pursue several – and often competing – purposes. Taxes on energy products may be set primarily to raise revenue, but they can also be used to integrate the costs of environmental and health damage into prices, so that energy users take these costs into account in their production and consumption decisions. Taxes and pricing policies may also be designed to keep energy prices low for reasons of equity to contain inflation or to stimulate economic growth. The weight given to these and possibly other motives in decisions related to energy taxes and pricing policy differs between countries due to their economic, social, and energy resource characteristics, and evolves over time.

Ideally, energy tax and pricing policies should aim to align end-user energy prices of energy with marginal production and social costs, as this helps the economy make the most productive use of resources while ensuring the well-being of the population and the environment in the most cost-efficient manner. Other policy objectives, including equity and economic development, are of course of strong importance in this discussion. However, economic analysis would suggest that where more targeted policy instruments exist for pursuing these objectives, these instruments should be used in preference to those that blunt or reduce the price signals provided in respect of energy products.

The role of price signals for energy use should also be considered in a dynamic sense, in terms of the signals they provide to long-term decisions that will influence future energy use patterns. There are advantages to bringing end-user energy prices more in line with production and environmental costs earlier rather than later. The reason is that decisions with difficult-to-reverse impacts (e.g. in relation to land use) or with long term impacts (e.g. investments in electricity generating plants or in transport infrastructure) are influenced by expectations of future energy prices, which in turn are affected by current prices and policies. Hence, if current prices accurately reflect production and societal costs, long-term decisions will support future energy use patterns with lower social costs, by encouraging more efficient or less-polluting energy use. Tax instruments therefore provide price signals that can help to modify the energy mix towards less harmful fuels, improve energy efficiency, or reduce demand for energy, both on a static and dynamic basis.

The uneven price signals with respect to different energy products, and the low tax rates that apply to many of them, suggest that with few exceptions the 41 countries considered do not harness the full power of taxes on energy use for environmental purposes. However, there is evidence that countries are progressing in this direction due to rising awareness of the negative side-effects of some sources of energy use and of the urgency of environmental problems related to some types of energy use. Among the selected partner economies, for example, China has introduced cap-and-trade systems that cover the second-largest amount of CO_2 emissions and is undertaking numerous measures to combat air pollution. Similarly, India is in the process of deregulating oil product prices, and has introduced a Clean Energy Cess on coal use, recently doubling the rate. In many selected partner economies, efforts are underway to stimulate the use of renewables in electricity generation, and across the 41 countries considered many are pursuing the G20 objective of phasing out inefficient support measures for fossil fuels. Such policies are indispensable to the promotion of sustainable development, which will require accommodating strong growth of energy use while containing the negative side effects. Taxes on energy are therefore an important part of the policy mix that can help to ensure that countries pursue their economic, social, and environmental objectives as effectively as possible.

Notes

1. Over the period considered, annual GDP has grown on average in all countries considered, so that considering changes over time approximates changes by GDP.
2. The USA is one of several countries with high per capita energy use and carbon emissions. Spain is a country of moderate energy intensity and carbon intensity of energy. In terms of energy use, like Spain, the Netherlands are a modal country but compared to Spain (and similar countries) it uses more energy per capita and energy is more carbon-intensive. The United Kingdom takes an intermediate position between Spain and the Netherlands.
3. The United States does not have a nationwide VAT. State-wide sales taxes are collected in 45 states, while 38 states collect local sales taxes. The five states with the highest average combined rates are Tennessee (9.45%), Louisiana (8.89%), Washington (8.88%) and Oklahoma (8.72%), while the five

states with the lowest average combined rates are Alaska (1.69%), Hawaii (4.35%), Wisconsin (5.43%), Wyoming (5.49%) and Maine (5.5 %).

4. The table excludes VAT concessions which may be related to energy use, but that are not directly related to fuels, such as VAT concessions for passenger transport or agriculture.
5. The EU VAT Directive requires a standard rate of at least 15% and reduced rates of at least 5%. The latter can only be applied to the set of goods and services listed in Annex III of the Directive. For more detail, see: *http://ec.europa.eu/taxation_customs/taxation/vat/how_vat_works/index_en.htm#vat_overview*.
6. Brazil also levies a higher rate of social security contributions on the sales revenue of energy products, translating into a higher specific taxation of energy products.
7. A detailed analysis of the distributional effects of consumption taxes can be found in OECD (2014).
8. Information on ETS is primarily taken from World Bank (2014).
9. The CO_2 emission figures have been derived from fuel use volumes using standard physical conversion factors from the sources set out in Annex A (see p. 212-13). This is possible since CO_2 emissions are generally fixed for given quantities of particular fuel types (subject to variations in fuel quality) regardless of the particular combustion technology used.

References

Arlinghaus, J. (2015), "Competitiveness Impacts of Carbon Pricing", *OECD Environment Working Paper*, OECD Publishing, Paris.

Copenhagen Economics (2007), "Study on Reduced VAT Applied to Goods and Services in the Member States of the European Union", *Final Report to the European Commission*, DG Taxation.

Dahl, D. (2012), "Measuring Global Gasoline and Diesel Price and Income Elasticities", *Energy Policy*, Vol. 41, pp. 2-13.

Energy Information Administration (several years), "Country Analysis Briefs". *www.eia.gov/countries/* (accessed October 2014).

Ernst and Young (2012), "The Worldwide VAT, GST and Sales Tax Guide", *2012 EYGM Ltd.*

European Commission (2014), "VAT Rates Applied in the Member States of the European Union, Situation at 1st July 2014." *European Commission*, Brussels.

G20 (2009), "Pittsburgh Summit Declaration", 24-25 September, *www.g20/org/*.

Harding, M. (2014), "The Diesel Differential: Differences in the Tax Treatment of Gasoline and Diesel for Road Use", *OECD Taxation Working Paper*, No. 21, OECD Publishing, Paris.

IEA (2014), "Extended World Energy Balances", *IEA World Energy Statistics and Balances* (database), Doi: *http://dx.doi.org/10.1787/data-00513-en*.

IPCC – Intergovernmental Panel on Climate Change (2006), "Guidelines for National Greenhouse Gas Inventories – Chapter 2", *www.ipcc-nggip.iges.or.jp/public/2006gl/vol2.html*.

Johansson, Å. et al. (2013), "Long-Term Growth Scenarios", *OECD Economics Department Working Papers*, No. 1 000, OECD Publishing.

Kojima, M. (2013), "Petroleum Product Pricing and Complementary Policies: Experience of 65 Developing Countries since 2009", *Policy Research Working Paper*, No. 6396, The World Bank, *http://elibrary.worldbank.org/doi/pdf/10.1596/1813-9450-6396*.

KPMG (several dates), "Country VAT/GST Essentials", *www.kpmg.com/Global/en/IssuesAndInsights/ArticlesPublications/vat-gst-essentials/Pages/default.aspx*.

Ministry of Economics and Public Finance Argentina (2012), "Estimación de los Gastos Tributarios Para los Años 2012 a 2014" (Estimation of Tax Expenditures for the years 2012 to 2014), *www.mecon.gov.ar/sip/dniaf/gastos_tributarios_2012-14.pdf*.

OECD (2015), "Inventory of Estimated Budgetary Support and Tax expenditures for Fossil Fuels 2015", OECD Publishing, Paris, forthcoming.

OECD (2009), "Declaration on Green Growth Adopted at the Meeting of the Council (MCM) at Ministerial Level on 25 June 2009", OECD, Paris, *www.oecd.org/officialdocuments/publicdisplaydocumentpdf/?doclanguage=enandcote=C/MIN(2009)5/ADD1/FINAL*.

OECD (2015), "Inventory of Estimated Budgetary Support and Tax expenditures for Fossil Fuels – A Report on Selected Non-Member Countries", *OECD Publishing*, Paris, forthcoming.

OECD (2014a), "Consumption Tax Trends", OECD Publishing, Paris, Doi: *http://dx.doi.org/10.1787/ctt-2014-en*.

OECD (2014b), "Database on Instruments Used for Environmental Policy, Environmentally Related Taxes", *www2.oecd.org/ecoinst/queries/*.

OECD (2014c), "The Distributional Effects of Consumption Taxes in OECD Countries", *OECD Tax Policy Studies*, No. 22.

OECD (2013a), "Inventory of Estimated Budgetary Support and Tax expenditures for Fossil Fuels", *OECD Publishing*, Paris, Doi: *http://dx.doi.org/10.1787/9789264187610-en*.

OECD (2013b), "Taxing Energy Use – A Graphical Analysis", *OECD Publishing*, Paris, Doi: *http://dx.doi.org/10.1787/9789264183933-en*

OECD (2013c), *Effective carbon prices*, OECD Publishing, Paris, Doi: *http://dx.doi.org/10.1787/9789264196964-en*.

OECD (2012), *Consumption Tax Trends 2012: VAT/GST and Excise Rates, Trends and Administration Issues*, OECD Publishing, Paris, Doi: *http://dx.doi.org/10.1787/ctt-2012-en*.

Parry, I. et al. (2014), "Getting Energy Prices Right – From Principle to Practice", *International Monetary Fund*, Washington, DC.

Ricardo – AEA (2014), "Update of the Handbook on External Costs of Transport", *Final Report for the European Commission – DG MOVE*, *http://ec.europa.eu/transport/themes/sustainable/studies/doc/2014-handbook-external-costs-transport.pdf*.

Wall Street Journal (2015), "Brazil Announces Tax Increases for 2015", *Wall Street Journal Latin America News*, 19 January, *www.wsj.com/articles/brazil-announces-tax-increases-for-2015-1421707755* (accessed January 2015).

World Bank (2014a), "State and Trends of Carbon Pricing 2014", *The World Bank*, Washington, DC.

World Bank (2014b), "World Development Indicators", The World Bank, Washington, DC.

Part II

Country profiles

Argentina

This chapter discusses the energy landscape, pricing policies, and the structure of energy taxation in Argentina and presents graphical profiles of energy use and taxation in Argentina in both energy and carbon terms (Figures 1.1 and 1.2, respectively).

Argentina is South America's largest natural gas producer, a significant producer of oil, and one of the world's largest producers of biodiesel. It has become increasingly dependent on energy imports, with the energy trade balance turning negative in 2011. Argentina continues to export crude oil and a very minor share of natural gas (0.07% in 2014), but most natural gas, liquefied natural gas (LNG), refined oil products and a small share of electricity are imported.

The Secretariat of Energy within the Ministry of Federal Planning, Public Investment and Services (MFPPIS) is responsible for national energy policy. Regulatory oversight is organised by sector: Between 2004 and 2014, ENARSA (Energía Argentina S.A.), itself a producer, was able to intervene in the oil market to prevent abuse of a dominant position by oil companies. In October 2014, ENARSA's exploration permits and concessions reverted back to the Secretary of Energy. Distribution and transport of natural gas are regulated by the National Gas Regulatory Authority (ENARGAS). Since 2013, the Commission for the Planning and Strategic Coordination of the National Hydrocarbon Investment Plan (CPSNHI) has gained increasing importance within the natural gas sector, due to its role in authorising natural gas projects which receive higher wellhead prices. The National Electricity Regulatory Authority (ENRE) monitors compliance with federal concession electricity contracts in the capital and Buenos Aires province, while other distributors are regulated by their respective provincial authorities. CAMMESA, the Wholesale Electricity Market Management Company (20% state-owned), manages the wholesale electricity market and the electricity grid (MFPPIS, 2014).

Oil and gas sector policies have undergone substantial change in recent years. Introduced in 2012, the Hydrocarbon Sovereignty Regime declares national self-sufficiency in fossil fuel supply to be a matter of public national interest and nationalised 51% of the oil company YPF and gas company Repsol YPF Gas (Ministry of Finance, 2012a and 2012b).

Exploration, production, transport and distribution of energy products are highly concentrated and vertically integrated. In the oil sector, YPF covers 41% of the market in exploration and extraction (the four largest companies cover 71%), and also accounts for 54% of refining (the four largest companies cover 66%). YPF also accounts for 28% of the market in natural gas extraction (the four largest companies share 75%), while transport and distribution are in the hands of regional monopolies (*ibid.*). While electricity generation is carried out in a mostly liberalised market, 95% of the electricity grid is managed by a single private company (Transener), and distribution is divided between regional monopolies (ADEERA, 2009; CAMMESA, 2012).

Large scale production of biodiesel from soybeans began in 2006 and bioethanol production from sugarcane or corn started in 2010. From 2014, all liquid fuel consumed in Argentina must be blended with a biodiesel or bioethanol content of 10% (Secretary of Energy, 2014). This proportion is expected to be increased further. The Program for Generation of Electricity through Renewable Sources (GENREN) was created to promote electricity generation from diverse renewable sources (Ministry of Foreign Affairs and

Worship, 2012). While hydroelectricity is still the main source of renewable energy (in 2009, 13% of electricity was made from hydropower), the country is currently expanding potential for electricity generation from wind, solar and geothermal energy (*ibid.*).

Main energy policies influencing producer prices

Oil and oil product prices are officially deregulated. In practice, the government has repeatedly required Argentina's five major oil companies to reduce prices between 2010 and 2012 (Kojima, 2013). Since 2012, fuel prices have been significantly increased and are now more closely linked to international oil price movements. Fuel prices are also centrally published by location and supplier, increasing transparency and incentivising companies to lower prices. Fuel tourism from Brazil is countered by charging higher prices to foreign cars at separate filling stations. Discounts for diesel for long-distance transport companies were discontinued in 2012 (Kojima, 2013). Companies pay royalties for natural gas and petroleum extraction to the respective provinces of up to 15%, except for offshore exploration, for which royalties are paid to the national state (WTO, 2013).

End-user prices of natural gas were fixed by ENARGAS in 2008 and were frozen until April 2014, despite demands by gas companies to increase prices (ENARGAS, 2014). In April 2014, the Secretary of Energy and ENARGAS more than doubled natural gas prices for all user categories to reduce the incidence of subsidies, but prices continue to lie below production cost. Rates vary by type of user and by regional distributor. Prices consist of a fixed and variable component, and consumers pay a billing charge that increases with use (*ibid.*). LPG and natural gas consumption in low-income areas and in areas with supply problems, including Patagonia, is subsidised via two different funds (Ministry of Finance, 2002). Gas fields exceeding projected production levels receive more than double the normal wellhead price, subject to approval by CPSNHI.

Except for some large users consuming electricity above a certain minimum consumption, electricity consumers buy electricity from regional monopolistic distributors at tariffs set by ENRE (CAMMESA, 2014). Electricity tariffs vary by user category and according to the amount of electricity consumed (ENRE, 2014), but, on average, users pay 23.5% of the cost of electricity generation (Caratori, 2014). Financed by a charge on each transaction in the electricity wholesale market, end-user electricity tariffs are reduced via the fund for regional End-User Tariff Compensation, which is administered at the provincial level, and are also strongly subsidised.

Biodiesel prices are set by MFPPIS for all producers. Since November 2013, prices have been differentiated by company size, with higher prices for small producers. Bioethanol prices are aligned with production costs.

Argentina levies a base duty of 5% on all exports, but duties on oil, natural gas, oil products, biodiesel and minerals are higher. Import tariffs are levied on Minerals and Metals with an average rate of 33.8% and a maximum rate of 35% (WTO, 2012). Annual quotas regulate imports of gas and diesel oil, which are collectively exempt from all specific taxes on fuel, but not from VAT. For 2012 and subsequent years, authorisation was granted for the import of 7 million m^3 of gasoline and 1 million m^3 of diesel (55%, and 12% of the domestic sales volume, respectively), but the imported fuel volume under this regime was less than half of the maximum volume authorised.

Structure of energy taxation

Argentina applies taxes to gasoline, diesel, fuel oil, kerosene, LPG, natural gas and, separately to compressed natural gas (CNG) for road use.[1] Other fuels, including other oil products, coal, and natural gas are untaxed. A tax on electricity is levied on each transaction on the electricity wholesale market.

The following taxes apply to oil and natural gas in Argentina:

- The **Tax on Liquid Fuels and Natural Gas (TLFN)** is levied on gasoline, diesel, fuel oil, kerosene and CNG for road use. The tax rates are expressed as *ad valorem* rates and also establish minimum rates expressed in per-unit values. A general exemption applies to all fuels consumed in the Patagonian region. Fuels used by agriculture, fisheries and mining, as well as commercial passenger transport are practically exempt, as they can credit fuel tax payments against other taxes. Biodiesel has been taxed under the TLFN since November 2011, when the TLFN exemption ended.
- The **Tax on Gasoline and Compressed Natural Gas**, applies to gasoline and natural gas, additional to the TLFN. Rates are expressed in *ad valorem* terms, with a higher rate for CNG, and minimum rates on per-unit terms for both fuels.
- The **Tax on Diesel and Liquefied Gas** applies to gasoil and liquefied gas for automotive use (additional to the TLFN) and is quoted in *ad valorem* terms.
- The **Surcharge on Natural Gas** is levied on the price of natural gas at ARS 0.004, charged at the point of entry into the distribution system. The surcharge is earmarked for a fund for residential gas consumption to subsidise LPG and natural gas tariffs in low-income areas or areas with supply problems.
- The **Surcharge on the Trust Fund for Imports of Natural Gas** is added to the price of natural gas, at between ARS 0.05 and ARS 0.27 per m³ of natural gas. Small residential users are exempt and rates are tiered by user category. Revenues feed into the Trust Fund for Imports of Natural Gas, used to finance imports of natural gas from Bolivia.
- The **Monitoring Fee of LPG Industry and Commercialisation** is levied on the refining of liquid fuels to LPG and LPG imports, with revenues feeding into a Fund subsidising the sale of liquefied petroleum gas in containers and cylinders.

Tax rates on these fuels are set out in Table 1.1. These taxes are included in the graphical profiles for Argentina.

Table 1.1. **Fuel taxes and charges on gasoline, CNG, diesel, LPG, kerosene and natural gas**

Rate as at 1 April 2012 (ARS per unit or % of sales price)	Gasoline (per litre)		CNG (road use, per m_3)	Diesel (per litre)	Fuel Oil (per litre)	Kerosene (per litre)	LPG	Natural gas (per m_3)		
	below 92 octanes	above 92 octanes						Ind.	Com.	Res.
Tax on liquid fuels and natural gas	70% (0.5375)	62% (0.5375)	16%	19% (0.15)	19% (0.15)	19% (0.15)	–	–	–	–
Tax on gasoline and CNG	5% (0.05)		9% (0.05/m³)	–	–	–	–	–	–	–
Tax on gasoil and liquefied gas	–		–	21%	–	–	–	–	–	–
Surcharge on natural gas	–		–	–	–	–	–	0.004	0.004	0.004
Surcharge on the trust Fund for imports of natural gas	–		–	–	–	–	–	0.0492	0.03	0.05
LPG monitoring fee	–		–	–	–	–	0.03	–	–	–
Total	75% (1.375)	67% (1.375)	25%	19% (0.15)	–	19% (0.15)	0.03	0.043	0.034	0.053

Source: Argentine Ministry of Finance (2013b).

Argentina also applies a tax to electricity when it enters the electricity wholesale market. The Surcharge for the National Electric Power Fund is charged upon each transaction on the wholesale market at ARS 0.0054676 per kWh. Revenues feed into the National Electric Power Fund for infrastructure and tariff subsidies. This surcharge is included in the graphical profiles for Argentina.

The VAT treatment of energy use in Argentina is summarised in Section 1.4.3. Natural gas and electricity are subject to a higher than standard VAT rate (27% instead of 21%), whereas LPG is subject to a lower rate (10.5%). As discussed in Section 2.2 of Part I, VAT is not shown in the graphical profiles for Argentina.

Energy use and taxation in Argentina

As can be seen in the graphical profiles for Argentina, transport accounts for about 19% of energy use in Argentina and 22% of carbon emissions. Energy use in transport is largely from diesel and gasoline, together accounting for 76% of energy use and 80% of carbon emissions. In this category, gasoline accounts for 33% of energy use and for about the same proportion of carbon emissions, while diesel accounts for 43% of transport energy and 46% of carbon emissions. A large proportion of transport energy use is natural gas (20% of transport energy), accounting for 16% of transport emissions. Most transport fuels are taxed under the Tax on Liquid Fuels and Natural Gas (TLFN). Gasoline and CNG are additionally taxed under the Tax on Gasoline and CNG, and Diesel under the Tax on Diesel and Liquefied Gas. Additional variation in tax rates is introduced when converting *ad valorem* tax rates into per-unit rates, due to differing prices among fuels. The highest cumulative tax applies to gasoline, at ARS 2.57 per litre, followed by diesel at ARS 1.66 per litre. Of aviation fuels, only kerosene is taxed, while aviation gas remains untaxed. While marine fuels account for a very small proportion of transport energy use (0.63%), of these, diesel and fuel oil are taxed under the TLFN at ARS 0.85 and ARS 0.74 per litre, respectively. CNG is cumulatively taxed at ARS 0.13 per litre; biodiesel is taxed at ARS 1 040 per ton since the end of the TLFN exemption for biodiesel.

Energy for heating and process use accounts for almost 50% of energy and 52% of carbon emissions. 62% of energy in this category is natural gas, accounting for about the same share of emissions, while oil products account for 27% of energy and 22% of carbon emissions. Within the heating and process category, kerosene is subject to the highest tax rate under the TLFN at ARS 0.81 per litre, while diesel and other oil products are not taxed. Natural gas is subject to the surcharge on natural gas and the surcharge for the Trust Fund for Natural Gas Consumption, with the latter differing according to user group. Cumulatively, residential and commercial users pay higher rates under the surcharge for the Trust Fund for Natural Gas Consumption and thus their natural gas consumption is taxed at ARS 0.054 per litre (res.), and ARS 0.034 per litre (com.). The surcharge for the Trust Fund of Natural Gas Consumption is lower for industrial users, thus natural gas consumption by industry is taxed at ARS 0.043 per litre. LPG is subject to the LPG Monitoring Fee (not visible in the graphical profiles).

Electricity is primarily generated from natural gas (accounting for 56% of energy and 68% of carbon emissions from this category), while renewables, combustibles and waste account for 13% of energy and 9% of emissions, and hydropower for 13% of energy with no emissions. Oil products and coal account for the remainder. The graphical profiles show the tax levied on electricity transactions in the wholesale market per kWh of electricity purchased as an implicit tax on the fuels used to generate electricity, using the "look-through approach" described in Section 2.2. In addition to this tax, natural gas for electricity generation is subject to the surcharge on natural gas for industrial users, as well as the charge for the Trust Fund for Natural Gas Consumption at ARS 0.043 per litre.

Reported tax expenditures and rebates from taxes shown in the graphical profiles

Tax expenditures are reported by the Ministry of Finance (2012d) in respect of the different tax treatment of gasoline, diesel and CNG under the TLFN, using the *ad valorem* rate levied on gasoline as a benchmark. These tax expenditures, using a benchmark for diesel and CNG that is the same as the *ad valorem* tax rate applied to gasoline, are shown in the graphical profiles.

Another tax expenditure in respect of the TLFN arises from the exemption for liquid fuels used in the South of the country in the Patagonia region. The differential tax treatment in the region is indicated in the graphical profiles using a +.

Under the Regime for the Sustainable Use and Production of Biofuels (Law 26.093), biofuels are exempt from the Tax on Liquid Fuels and Natural Gas as well as from the Tax on Naphtha and CNG. Following the removal of the exemption for biodiesel in 2011, tax expenditures from the Tax on Liquid Fuels and Natural Gas arise only with respect to the tax treatment of bioethanol. Since no bioethanol was used in 2009 (the year for which energy use is shown in the graphical profiles) this tax expenditure cannot be shown in the graphical profiles.

Key assumptions and caveats

Key assumptions and caveats are as follows:

- All natural gas for road use in the transport category has been assumed to be compressed natural gas.
- The price for fuel oil has been assumed to be the price for the lowest diesel grade.

Price information has been retrieved from Secretary of Energy (2014). Prices used to calculate *ad valorem* tax rates are set out in Table 1.2.

Table 1.2. **Price information used to convert ad valorem rates to per unit rates in Argentina**

Gasoline	ARS 3.83 per litre, which is the weighted average of price of gasoline grade 1 and grades 2 and 3
Diesel	ARS 4.05 per litre, which is the simple average price of diesel grades 1, 2, 2B and 3.
Kerosene	ARS 4.28 per litre
Fuel oil	ARS 3.89 per litre (price of grade 1 diesel)
Compressed natural gas	ARS 1.48 per litre

Information on tax rates and bases, tax expenditures energy use, and conversion rates were obtained from the sources detailed in Annex A, from Government websites, and from consultation with national officials.

In addition, country-specific sources were used as listed in the references for this chapter.

Figure 1.1. **Taxation of energy in Argentina on an energy content basis**

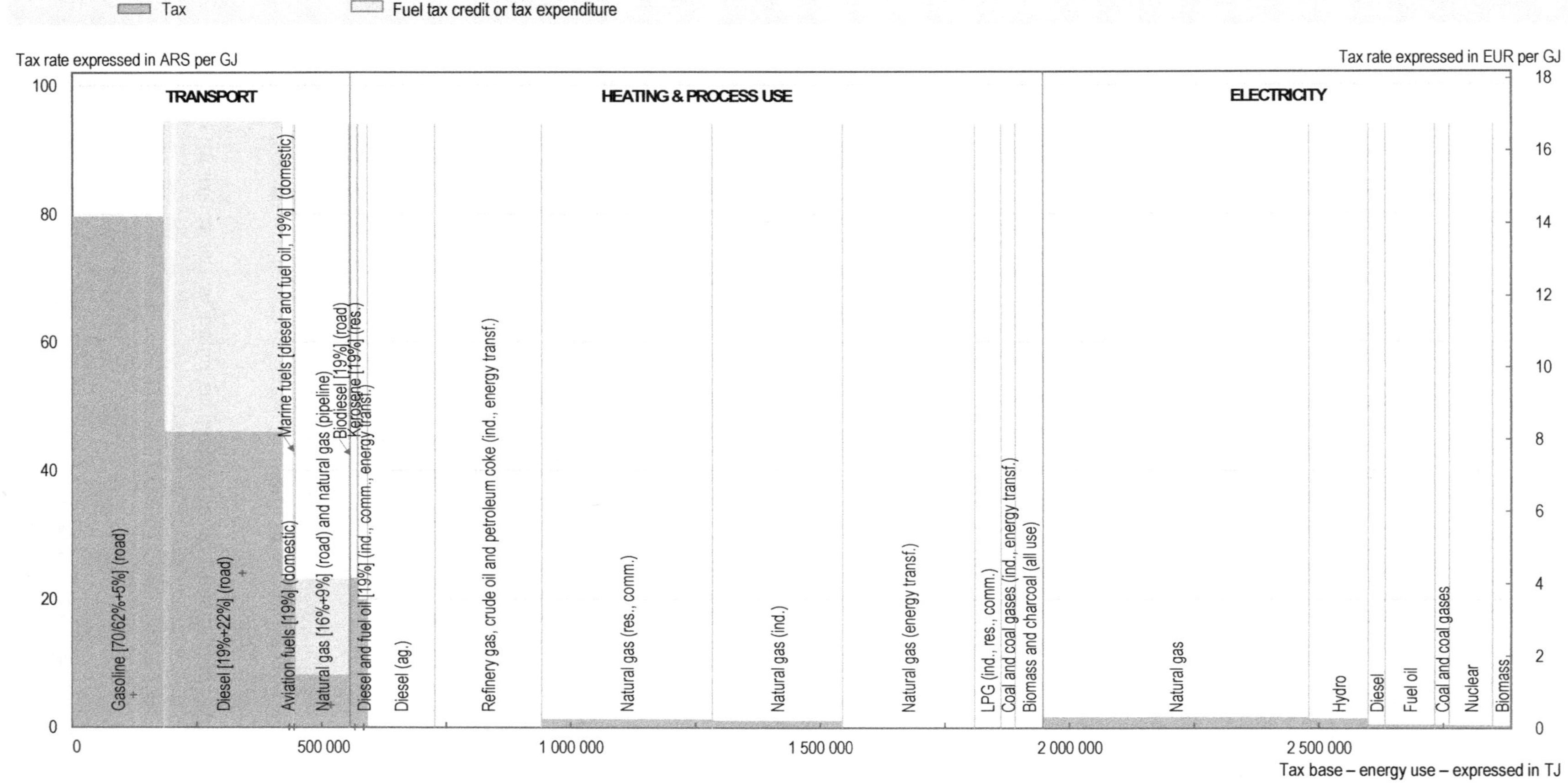

Source: OECD calculations. Tax rates are as of 1 April 2012; energy use data is for 2009 from IEA (2014), *IEA World Energy Statistics and Balances* (database), Doi: http://dx.doi.org/10.1787/data-00513-en. The exemption from the fuel tax in Patagonia is indicated in the graphical profiles using a +. Tax rates on gasoline (road), diesel (road), aviation and marine fuels, natural gas (road) and biodiesel (road) have been calculated as a percentage of their price, subject to the caveats described in this chapter.

StatLink http://dx.doi.org/10.1787/888933205809

Figure 1.2. **Taxation of energy in Argentina on a carbon content basis**

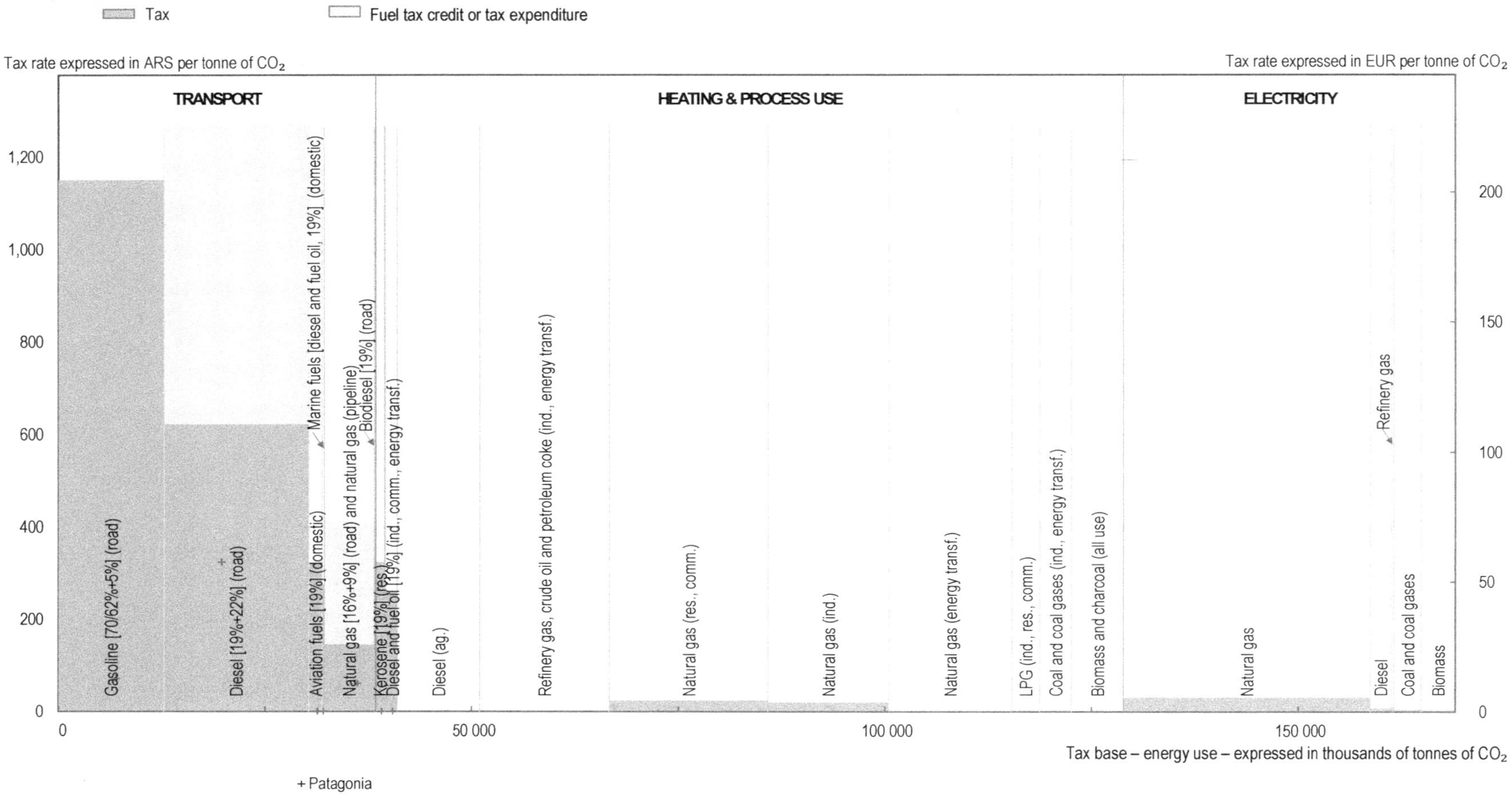

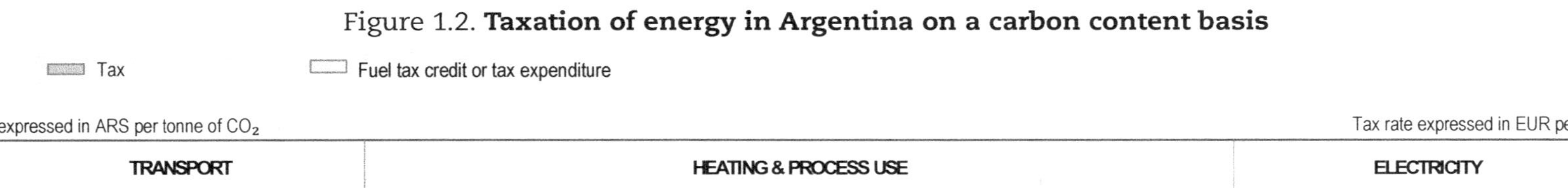

Source: OECD calculations. Tax rates are as of 1 April 2012; energy use data is for 2009 from IEA (2014), *IEA World Energy Statistics and Balances* (database), Doi: http://dx.doi.org/10.1787/data-00513-en. The exemption from the fuel tax in Patagonia is indicated in the graphical profiles using a +. Tax rates on gasoline (road), diesel (road), aviation and marine fuels, natural gas (road) and biodiesel (road) have been calculated as a percentage of their price, subject to the caveats described in this chapter.

StatLink http://dx.doi.org/10.1787/888933205817

Note

1. Most information on tax rates in Argentina has been retrieved from Ministry of Finance (2013b).

References

ADEERA (2009), "Datos Característicos de las Empresas, 2009", *www.adeera.com.ar/* (accessed October 2014).

CAMMESA (2012), "Informe Annual", available at: *http://portalweb.cammesa.com/memnet1/Pages/descargas.aspx* (accessed October 2014).

Caratori, L. (2014), "Las Inverosimilitudes Energéticas del Presupuesto 2015", Instituto Argentino de la Energía "General Mosconi", *http://web.iae.org.ar/recomendados/las-inverosimilitudes-energeticas-del-presupuesto-2015* (accessed December 2014).

ENARGAS (2008), "Resolución 409/2008 – Establécese la Segmentación de las Categorías Definidas en el Decreto No. 181/04 Respecto de los Usuarios Residenciales", *http://mepriv.mecon.gov.ar/Normas2/409-08.htm* (accessed October 2014).

ENARGAS (1992), "Marco Regulatorio de la Industria de Gas, Ley No. 24076", *www.enargas.gov.ar/MarcoLegal/Ley_24076.pdf* (accessed October 2014).

ENRE National Electricity Regulator (2014), "Sobre el Calcúlo Tarifario" (On tariff calculation), *www.enre.gov.ar/* (accessed October 2014).

IEA (2014), "Extended World Energy Balances", *IEA World Energy Statistics and Balances* (database), Doi: *http://www.dx.doi.com/10.1787/data-00513-en.*

Ministry of Economics and Public Finance (2013a), "Fondo Fiduciario del Sistema de Infraestructura Hídrica (Decreto No. 1381/01)", *www.ucofin.gov.ar/pdf/consumo%20nafta.pdf.*

Ministry of Economics and Public Finance (2013b), "Tributos Vigentes en la República Argentina a Nivel Nacional" (Taxes in the Argentinean Republic at National Level), Dirección Nacional de Investigaciones y Analisis Fical, Secretaria de Hacienda, *www.mecon.gov.ar/sip/dniaf/tributos_vigentes.pdf* (accessed October 2014).

Ministry of Economics and Public Finance (2012a), "Reglamento del Régimen de Soberanía Hidrocarburífera de la República Argentina, Ley 26.741/2012" (Regulation of the Hydrocarbon Sovereignty Regime of the Argentinean Republic), *http://mepriv.mecon.gov.ar/Normas2/26741.htm.*

Ministry of Economics and Public Finance (2012b), "Soberanía Hidrocarburifera (Decreto 1277/2012" (Hydrocarbon Sovereignty Regime 1277/2012), *http://infoleg.mecon.gov.ar/infolegInternet/anexos/200000-204999/200130/norma.htm.*

Ministry of Economics and Public Finance (2012c), "Estimación de los Gastos Tributarios para los Años 2012 a 2014" (Estimation of Tax Expenditures for the years 2012 to 2014), *www.mecon.gov.ar/sip/dniaf/gastos_tributarios_2012-14.pdf.*

Ministry of Economics and Public Finance (2011), "Complejo Petróleo y Gas, Serie Produccion Regional por Complejos Productivos" (Petroleum and Gas Complex, Series Regional Production by Productive Complexes), Secretaría de Política Económica, Subsecretaría de Progamación Económica, *www.mecon.gov.ar/peconomica/docs/Complejo_Petroleo_y_Gas.pdf.*

Ministry of Foreign Affairs and Worship (2012), "Fuelling Development – Renewable Energies in Argentina", *www.inversiones.gov.ar/userfiles/folleto_energias_renovables_-_ingles.pdf.*

Secretary of Energy (2014), Consulta de Precios de EESS de la Resolución S.E. 1104/2004, *http://res1104.se.gov.ar/consultaprecios.eess.php* (accessed October 2014).

Brazil

This chapter discusses the energy landscape, pricing policies, and the structure of energy taxation in Brazil and presents graphical profiles of energy use and taxation in Brazil in both energy and carbon terms (Figures 2.1 and 2.2, respectively).

Brazil is Latin America's second-largest holder of reserves and second-largest producer of oil and natural gas. It is also the continent's largest oil and second-largest natural gas consumer (BP Statistical Review of World Energy, 2014). It exports crude oil but imports growing amounts of coal, petroleum products, natural gas and LNG. With one-fourth of global production in 2013, Brazil is the world's second largest biofuel producer and consumer. Although Brazil produces some coal, 98% of coal is imported.

The Ministry of Mines and Energy (MME) sets the framework for energy policy. Under the MME, the Autonomous National Agency for Petroleum, Natural Gas and Biofuels (ANP) is the main government agency charged with implementing national policy for oil, natural gas (except for the regulation of natural gas distribution at state-level) and, since 2011, biofuels (ANP, 2014; Gomes, 2014). The National Agency for Electric Energy (ANEEL) regulates the electricity sector.

The discovery of pre-salt oil and gas changed the Brazilian landscape for fossil fuel exploration, contributing to a 19% growth in oil reserves and a 26% growth in those of natural gas between 2007 and 2011. The government created a state-owned enterprise, Pré-Sal Petroleo S.A., to manage its interests in the exploration and production of the pre-salt reserves and gave some exclusive rights to government-owned Petrobras. Although Petrobras' monopoly on fossil fuel exploration, production, refining and imports was abolished in 1997, Petrobras continues to dominate oil exploration and production (91%), refining (98%), distribution (38.1%) and service stations (20%) (Kojima, 2013; ANP, 2013). The 2009 Gas Law aimed to unbundle the gas industry, but has not yet been implemented (Mutsaers, 2014). Petrobras also controls 90% of natural gas reserves and production, transmission and imports, while private producers have market shares not exceeding 5% each. While almost 70% of the distribution network is operated by a Petrobras-subsidiary, Petrobras holds shares of most remaining pipeline operators (WTO, 2013). Natural gas distribution and service station ownership are more diversified.

In the electricity sector, Eletrobras controls around 40% of generation capacity. Other generation companies, in subnational or private hands, all hold market shares below 10% each. While most of the grid is publicly managed, 60% of distribution capacity is privately held (Brazilian Association of Electricity Distributors, n.d.).

Due to its comparatively early decision to support renewable energy, Brazil's share of renewable energy in total primary energy demand is far above the global average (IEA, 2013). More than 70% of Brazil's electricity is generated from hydropower, and in the transport sector, more than 15% of consumption is from domestically produced biofuels (*ibid.*). Biodiesel production is fragmented and primarily carried out by private firms, although Petrobras is the largest producer (about 25%), and is also the sole biodiesel buyer (ANP, n.d.). Ethanol production is highly fragmented, with Petrobras being the third largest producer (ANP, 2014). Regulation requires producers to blend all diesel with biodiesel at a proportion of 5 to 7%, and gasoline to be blended with ethanol, at a 25% rate.

Main energy policies affecting producer prices

Oil prices were deregulated in 2002, but in practice were determined by Petrobras until they were frozen by the government in 2006. Low ex-refinery gasoline prices increased fuel demand after the price freeze, obliging Petrobras to sell imported fuels purchased at world prices at a loss. Fuel prices were raised twice in 2012, and in January and February 2013. The CIDE tax on fuels was temporarily set to zero simultaneously with the June 2012 price changes (and was re-instated in February 2015). Bioethanol consumption is inversely related to gasoline prices, as flex-fuel vehicles are able to choose the cheapest option. In 2009, bioethanol sales exceeded gasoline sales but have since then decreased in line with lower gasoline prices.

Natural gas producers negotiate gas supply prices with gas distribution companies and captive consumers that consume gas above a certain threshold. In the downstream segment, 27 regionally monopolistic gas distribution companies supply gas to consumers at prices set by state regulatory agencies. Prices of domestically-produced natural gas are generally about one-third higher than those of natural gas imports. The dry season, and the associated lack of hydropower, in 2012 and 2013 prompted Petrobras to import LNG at Asian-equivalent spot prices to meet the needs of its gas-fired power plants.

The electricity market has both a regulated and an unregulated segment. The regulated market (75% of electricity consumption) covers all distribution companies on the national electricity grid, which deliver electricity to their assigned residential consumers at prices defined by ANEEL. Prices are tiered by user category and vary by region. In the unregulated market segment, large industrial consumers (above 3 MW) acquire electricity directly from generation companies and negotiate long-term supply contracts (about 25% of the market). To lower the relatively high electricity prices for Brazilian industry and to combat inflation, the government announced a range of electricity tax cuts in September 2012, reducing retail prices by 20% (Rhodes, 2012). By accepting substantially lower prices, concession holders were able to avoid new tenders upon the expiration of their contracts in 2013 and 2014 (OECD, 2013).

A number of programmes subsidise electricity consumption. In the effort to decrease reliance on hydropower, the Thermoelectric Priority Programme established that gas-fired thermoelectric power plants receive gas at a regulated maximum price, which is set at less than half the price of nationally produced natural gas and at about two-thirds of the price of imported gas. Three funds (the Global Reversion Fund, the Fuel Consumption Fund and the Energy Development Fund) with overlapping objectives are financed by charges on electricity, some of which were reformed in September 2012. Their joint purpose is to finance rural electrification, to compensate power plants for the high cost of electricity generation in the north of the country and to support low-income households.

Some other taxes feature specific regimes for energy products, but are not shown in the graphical profiles. Brazilian social security contributions (PIS/COFINS) are paid by companies on sales revenues. Most petroleum products are subject to a special PIS and COFINS regime (PIS/COFINS Monofásico), levying specific, higher *ad valorem* rates on the sale of gasoline, diesel, LPG, aviation fuel, biodiesel, and lower rates on naphtha. However, in practice most taxpayers choose to pay the PIS/COFINS tax under the *ad rem* regime, which levies a lower effective tax rate, A number of tax expenditures arise from exemption of energy products from PIS/COFINS. The REPEX programme allows oil companies to benefit from PIS/COFINS exemptions and some other taxes for their imports of petroleum products. PIS/COFINS taxes are also reduced for the import and retail of naphtha, and coal and natural gas-fired thermoelectric power plants are exempt for natural gas purchase.

A tax of between 30% and 150% of the sales price can be levied on exports, but is practically zero-rated for most products. Imports are subject to an import duty of up to 35%, in addition to two other taxes – the Imposto sobre Productos Industrializados (IPI) (from which energy products are exempt) and the Imposto sobre Circulação de Mercadorias e Serviços (ICMS). An import tariff of 14% applies to biodiesel and there is a temporary waiver of the 20% import duty on ethanol.

Structure of energy taxation

Taxes in Brazil are levied by the federal government, state governments and municipalities. The federal fuels tax (Contribution for Intervention in the Economic Domain on fuels, CIDE) is paid by fuel producers and importers on the import and sale of oil and oil products, natural gas and derivatives and bioethanol. Exemptions apply to exports and naphtha aimed at developing petrochemicals and tax revenue is earmarked for subsidising fuel, natural gas and oil product prices, for the financing of environmental projects and for transport infrastructure programmes.

CIDE rates are set by government decree. The CIDE tax is *de facto* used by the government to compensate for fuel price variations. In response to a diesel and gasoline price increase the government set CIDE tax rates on gasoline and diesel to zero in June 2012 (Bloomberg, 2012; Decree 7.764/2012). The CIDE rates for import and sale of kerosene, fuel oil, LPG from natural gas and naphtha, ethanol fuel had already been set to zero in April 2004. The CIDE tax and associated tax expenditures are shown in the graphical profiles. At the beginning of 2015, the CIDE tax was re-instated.

Two charges are levied on electricity consumption, which are both shown in the graphical profiles:

- The **Fuel Consumption Charge** (Quota de Consumo de Combustiveis, CCC) is levied on electricity distribution and transmission companies and passed on to end customers, who pay the charge as part of their electricity bill (ANEEL, 2013). While this charge is shown in the graphical profiles at BRL 7.75 per MWh, it was discontinued in September 2012. CCC revenues flow into the Fuel Consumption Fund.
- The **Energy Development Charge** (Conta de Desenvolvimento Energetico, CDE) is levied on electricity distribution and transmission companies and feeds into the Energy Development Fund. The charge is set at BRL 0.65 per MWh in the North and North-East and BRL 2.95 per MWh in the South and South-East. Self- and independent producers, as well as some small consumers, are exempt from the CDE, and low-income consumers pay discounted rates. Funds are allocated by ANEEL to the distribution companies to ensure reduced tariffs for poor families (OECD, 2008). In the graphical profiles, the CDE charge is shown as a weighted average for the northern and southern regions, as at 1 July 2012. However, the CDE charge was cut by 73% in the course of the September 2012 reform.

Brazil's VAT policy is summarized in Section 1.4.3. In Brazil, state VAT is levied at a rate of 17 to 18%, with lower rates levied on natural gas purchases (12%) and a higher rate for electricity (25%). As discussed in Section 2.2, VAT is not shown in the graphical profiles for Brazil.

Energy use and taxation

As displayed in the graphical profiles for Brazil, transport makes up 28% of total energy use and accounts for about the same amount of carbon emissions. Transport is dominated by gasoline (24% of transport energy and 22% of carbon emissions) and diesel (47% of transport energy and 46% of emissions). Biodiesel and other bio-liquids account for 16% of energy used (17% of emissions) in this category, whereas bioethanol for 5% of energy and emissions. The remainder is made up of natural gas and other oil products. Due to the temporary suspension of the CIDE tax, which is levied on gasoline and diesel, transport fuels were *de facto* untaxed (zero-rated) from June 2012. As mentioned above, the CIDE tax was re-instated at the beginning of 2015.

Energy for heating and process use accounts for 49% of total energy. Within this category, oil products account for 28% of energy and 22% of emissions, while combustibles and waste are the largest fuel, accounting for 53% of energy and 63% of emissions. Natural gas, coal and solar energy account for the rest. Between the suspension of the CIDE tax in June 2012 and its re-introduction in 2015, diesel was *de facto* untaxed (zero-rated). All other fuels in this category are untaxed.

Electricity is largely generated from hydropower, which accounts for 68% of energy used to generate electricity. The second largest source of energy for electricity generation is combustibles and waste, accounting for 9% of energy and 40% of emissions from this category. Other fuels used for electricity generation are oil products, coal, natural gas and a small amount of wind energy. The CIDE tax applies to fuel inputs to electricity generation. Having set rates to zero in 2012, this results in a tax expenditure from the CIDE tax for diesel inputs to electricity generation. Through the Fuel Consumption Fund and the Energy Development Charge, Brazil also taxes electricity consumption. While the Fuel Consumption Charge is levied throughout the country at a fixed rate, the Energy Development Charge differs between the northern and southern regions. The graphical profiles show the full amount of the Fuel Consumption Charge and adds to this a weighted average Energy Development Charge (based on consumption), indicating the two rates that apply in the northern and southern regions with + and ▪ respectively. The graphical profiles show the combined rates from these taxes as implicit tax rates on the underlying fuels used in electricity generation, using the look-through approach described in Section 2.2.

Tax expenditures and rebates relevant for the graphical profiles

Between June 2012 and February 2015, the CIDE tax rates on imports and retail of gasoline and diesel were set to zero in order to offset increases in the ex-refinery prices of petroleum products. This tax expenditure is shown in the graphical profiles in the categories "gasoline (road)" and "diesel (road)".

The Fuel Consumption Fund reimburses the difference between the northern region's high cost of electricity generation and the national average cost. The Energy Development Fund, partly financed by a charge on electricity distribution companies, supports five diesel and coal-fired power plants. Both of these tax expenditures are not shown in the graphical profiles.

Key assumptions and caveats

Key assumptions and caveats are as follows:

- As a consequence of major changes of the CIDE tax on fuels made in June 2012, the graphical profiles show the updated rates as of 1 July 2012.
- Though shown in the graphical profiles, the charge on electricity feeding into the Fuel Consumption Account was discontinued in September 2012 and the Energy Development Charge was reduced. While electricity is still taxed in Brazil, the 2012 tax reform substantially reduces the level of electricity taxes shown in the graphical profiles.

Information on tax rates and bases, tax expenditures energy use, and conversion rates were obtained from the sources detailed in Annex A, from Government websites, and from consultation with national officials.

In addition, country-specific sources were used as listed in the references for this chapter.

Figure 2.1. **Taxation of energy in Brazil on an energy content basis**

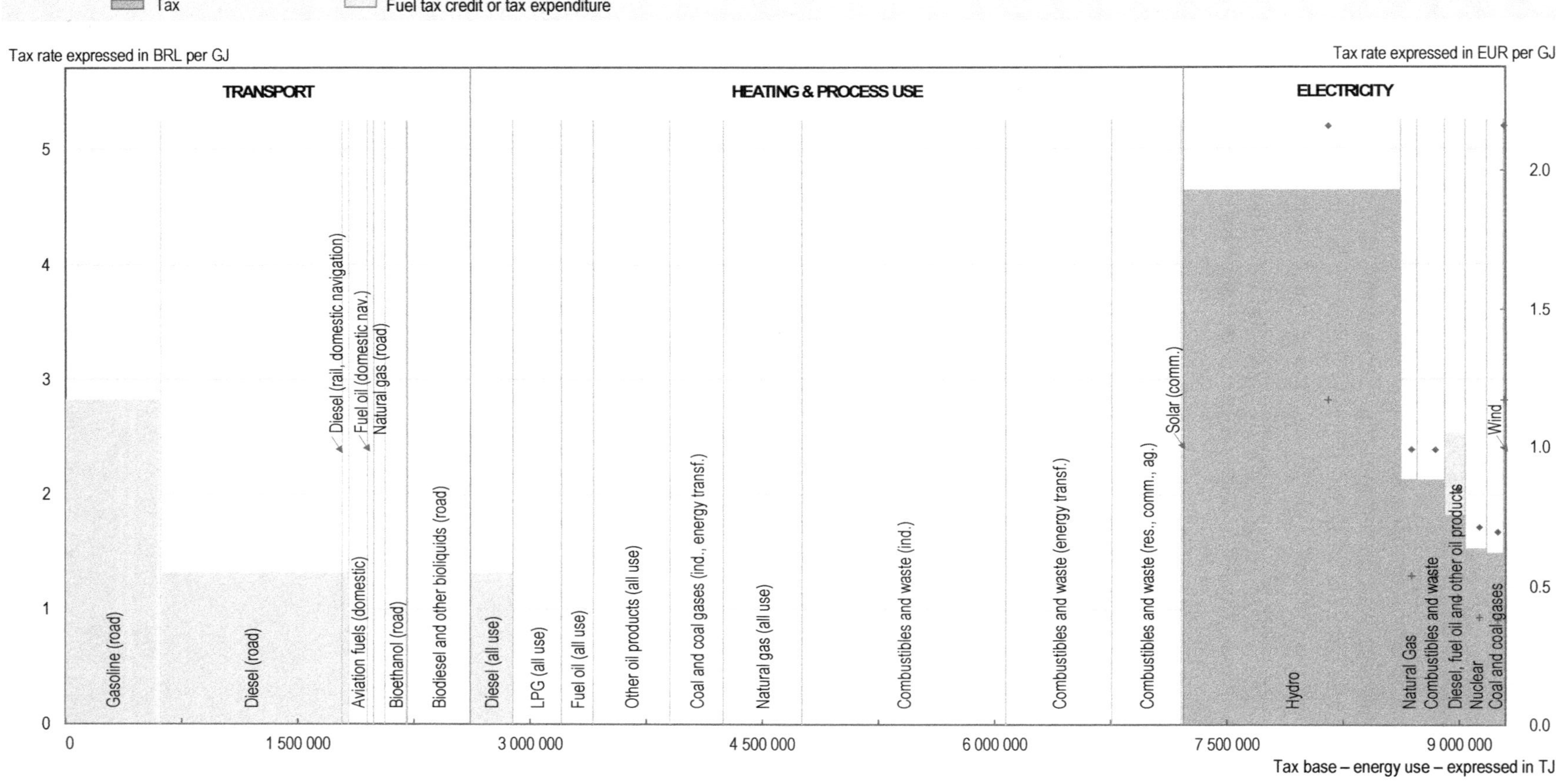

Source: OECD calculations. Tax rates are as of 1 July 2012; energy use data is for 2009 from IEA (2011a). Tax rates in the electricity sector shown in the dark grey shaded area reflect the Fuel Consumption Charge together with a weighted average rate for the Energy Development Charge across the North and South. The level of rates in the North are indicated by a + and the level of the tax rates in the South are indicated by a ■.

StatLink *http://dx.doi.org/10.1787/888933205827*

Figure 2.2. **Taxation of energy in Brazil on a carbon content basis**

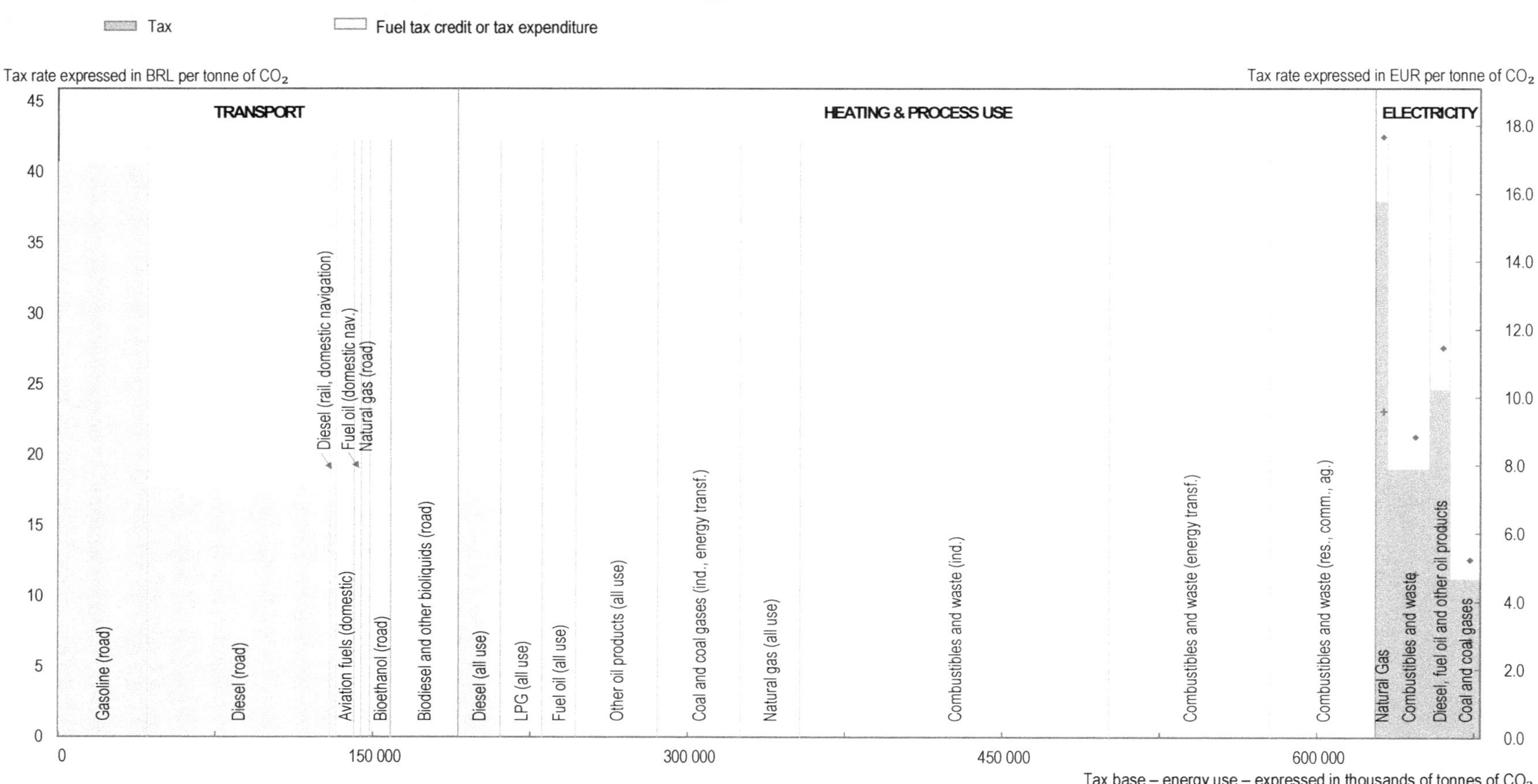

Source: OECD calculations. Tax rates are as of 1 July 2012; energy use data is for 2009 from IEA (2011). Tax rates in the electricity sector shown in the dark grey shaded area reflect the weighted average rates across the North and South. The level of rates in the North are indicated by a + and the level of the tax rates in the South are indicated by a ■.

StatLink http://dx.doi.org/10.1787/888933205834

References

ANEEL (2013), "CCC's Annual Budget to be Discussed at Public Hearing", *www.aneel.gov.br/aplicacoes/noticias_area/arquivo.cfm?tipo=PDFandidNoticia=6387andidAreaNoticia=347* (accessed October 2014).

ANP (2013), "Anuário Estatístico Brasileiro do Petróleo, Gás Natural e Biocombustíveis", *www.anp.gov.br/*.

ANP (2002), "Portaria Interministerial No. 234, 2002", *http://nxt.anp.gov.br/NXT/gateway.dll/leg/folder_portarias/portarias_interm/2002/pinterm%20234%20-%202002.xml* (accessed October 2014).

ANP (n.d.), "Natural Gas Tariffs and Prices", *www.anp.gov.br/?pg=65637andm=andt1=andt2=andt3=andt4=andar=andps=andcachebust=1403886498451* (accessed October 2014).

ANP (n.d.), "Preços de Produtores e Importadores de Derivados de Petróleo", *www.anp.gov.br/?pg=45148* (accessed October 2014).

ANP (n.d.), "Participações Governamentais e de Terceiros na Lei do Petróleo", *www.anp.gov.br/brnd/round10/arquivos/seminarios/SJF_Participacoes_Governamentais.pdf* (accessed October 2014).

Bloomberg (2012), "Petrobras to Raise Gasoline Prices as Brazil Cuts Tax", 22 June 2012, *www.businessweek.com/news/2012-06-22/petrobras-to-raise-gasoline-prices-as-brazil-cuts-tax*.

Brazilian Federal Government (2013), "Lei 12783/1", *www.planalto.gov.br/ccivil_03/_ato2011-2014/2013/lei/L12783.htm*.

Brazilian Federal Government (2011), "Law 12.431 of June 2011", *www.planalto.gov.br/ccivil_03/_Ato2011-2014/2011/Lei/L12431.htm*.

Brazilian Federal Government (1997), "Petroleum Law", Cabinet for Legal Affairs, Brasilia, *www.planalto.gov.br/ccivil_03/leis/l9478.htm*.

Brazilian Federal Government (2000), "Programa Prioritário de Termeletricidade", *www.planalto.gov.br/ccivil_03/decreto/D3371.htm*.

Departamento Nacional de Producao Mineral (n.d.), "Compensação Financeira pela Exploração de Recursos Minerais", *www.dnpm.gov.br/conteudo.asp?IDSecao=60*.

Economist Intelligence Unit (2013), "Petrobras Urges Government to Lift Petrol Prices", *www.eiu.com/industry/article/1440917528/brazil-oil-quick-view--petrobras-urges-government-to-lift-petrol-prices/2013-09-03*.

Eletrobras (2010a), "Fundos Setoriais", *www.eletrobras.com/ELB/data/Pages/LUMISB29596DDPTBRIE.htm*.

Firjan (2011), "Tarifa de Energia Para Indústria Brasileira Está 50% Acima da Média Mundial", *www.firjan.org.br/data/pages/2C908CEC30E85C950131B3B6A4A069BE.htm*.

Gomes, I. (2014), "Brazil: Country of the Future or Has its Time Come for Natural Gas?" Oxford Institute for Energy Studies, NG 88, June, available at: *www.oxfordenergy.org/wpcms/wp-content/uploads/2014/07/NG-88.pdf*.

IEA (2014), "Extended World Energy Balances", *IEA World Energy Statistics and Balances* (database), Doi: *http://dx.doi.org/10.1787/data-00513-en*.

IEA (2013), "World Energy Outlook 2013". *OECD/IEA Publishing*, Paris.

Ministry of Energy and Mines (2012), "Concessions for Electricity Generation, Transmission and Distributions – Questions and Answers", *www.mme.gov.br/mme/galerias/arquivos/noticias/2012/Perguntas_e_respostas_-_Concessxes_10_09_12_final.pdf*.

Mutsaers (2014), "Brazilian Government to Launch First Natural Gas Transportation Tender", *www.mondaq.co.uk/x/300126/Oil+Gas+Electricity/Brazilian+government+to+launch+first+natural+gas+transportation+tender*.

OECD (2008), *OECD Reviews of Regulatory Reform: Brazil 2008: Strengthening Governance for Growth*, OECD Publishing, Paris, Doi: *http://dx.doi.org/10.1787/9789264042940-en*.

Operador Nacional do Sistema Eléctrico (n.d.), "Energia Natural Afluente", *www.ons.org.br/historico/energia_natural_afluente_out.aspx*.

Rhodes (2012), "Brazilian Electricity Market Reforms", *www.mondaq.com/x/197544/Oil+Gas+Electricity/Brazilian+Electricity+Market+Reform* (accessed October 2014).

Receita Federal (2006), "IPI Tax Incidence – Aprova a Tabela de Incidência do Imposto Sobre Produtos Industrializados", *www.receita.fazenda.gov.br/Legislacao/Decretos/2006/dec6006.htm*.

Receita Federal (2001), "DCide, Lei 10.336 de 19/12/2001", *www.receita.fazenda.gov.br/pessoajuridica/cidecomb/default.htm*.

Receita Federal (n.d.), "Regime de Incidência Não-Cumulativa PIS e Cofins", *www.receita.fazenda.gov.br/PessoaJuridica/PisPasepCofins/RegIncidencia.htm*.

Reuters (2014), Brazil Raises Biodiesel Requirement in Diesel Blend", *www.reuters.com/article/2014/05/28/brazil-biodiesel-idUSL1N0OE0S220140528*.

Wall Street Journal (2015), "Brazil Announces Tax Increases for 2015", *Wall Street Journal Latin America News*, 19 January 2015, *www.wsj.com/articles/brazil-announces-tax-increases-for-2015-1421707755* (accessed January 2015).

Woolf et al. (2010), "Brazil's Electricity Market: A Successful Journey and an Interesting Destination", *www.mondaq.com/x/93780/Oil+Gas+Electricity/Brazils+Electricity+Market+A+Successful+Journey+And+An+Interesting+Destination*.

People's Republic of China

This chapter discusses the energy landscape, pricing policies, and the structure of energy taxation in China and presents graphical profiles of energy use and taxation in China in both energy and carbon terms (Figures 3.1 and 3.2, respectively).

China's fast-growing economy has led it to be the world's largest energy consumer and producer. Its increasing demand for oil imports puts upward pressure on world oil prices. China is the world's largest producer, consumer and importer of coal, and coal continues to dominate China's energy supply. Over the past decade, the country has increasingly relied on crude oil and natural gas imports.

Publicly-owned energy companies are overseen by the State-owned Assets Supervision and Administration Commission (SASAC). The primary regulatory body in the energy sector is the National Development and Reform Commission (NDRC). Created in 2008, the National Energy Agency (NEA), which operates under the NDRC, sets domestic wholesale energy prices and implements energy policies. Both SASAC and NDRC operate directly under the chief administrative authority of China, the State Council.

Ownership of the oil, natural gas and electricity sectors is highly concentrated among state-owned companies. The oil and natural gas sectors are dominated by China National Petroleum Corporation (CNPC) and China Petroleum and Chemical Corporation (Sinopec), both State-owned. CNPC is the leading player in oil exploration and production, along with its publicly-listed arm, Petro China and they account for 53% and 75% of total oil and natural gas output, respectively (EIA, 2014). Whereas Sinopec focuses on refining and distribution, CNPC has been expanding its presence in these markets over the past years and new state-owned firms have emerged, too. CNPC is also the country's largest natural gas company. The state-owned China National Offshore Oil Corporation (CNOOC), the third largest oil company, is responsible for offshore oil exploration and production and has more recently started activities in refining (*ibid.*).

Although the past decade has seen some consolidation, China's coal industry remains fragmented and primarily public. The ten largest companies produced over a third of domestic coal in 2011, with local, town and village mines accounting for the remainder. Foreign investment is increasingly sought to help to modernise existing large-scale mines and to introduce new technologies (EIA, 2014).

Five state-owned electricity companies generate about half of total electricity. Independent power producers generate the remainder, often in partnership with privately-listed subsidiaries of the state-owned companies. Electricity transmission and distribution is managed by two State-owned companies, the Southern Power Grid Company and the State Power Grid Company.

China is the world leader in renewable energy investment, and in 2013, China invested more in renewable energy than Europe (UNEP, 2014). Though China's USD 56 billion of investment shrunk by 6% between 2012 and 2013, it was still more than Europe's USD 48 billion, which reduced by 44% (*ibid.*). China's installed wind capacity has risen from virtually none in the early 2000s to the world's largest source of wind generation. Similarly, China accounted for the majority of the global expansion of installed solar water heating capacity during the 2000s and currently has half of the world's electricity generation capacity from solar (OECD, 2013). Feed-in tariffs for renewable energy have supported renewable energy investment – for solar energy, the tariff is more than double the amount paid for electricity from coal, and biomass projects receive a 50% premium (UNEP, 2013). The feed-in tariff for wind has been declining, and is now similar to that of other energy sources. Despite recent and ongoing growth in renewables, the share of renewables in total energy production in 2011 was under 10%, and consisted primarily of hydropower (OECD, 2013). In November 2014, China announced a target to increase its share of renewable energy in total energy use to 20% by 2030, and to have its CO_2 emissions peak in the same year.

Main energy policies influencing producer prices

The NDRC, through the NEA, regulates the prices of gasoline, diesel, and natural gas, at both the wholesale and retail levels. Upon NDRC's communication of benchmark prices for each regional capital, provincial authorities issue price caps for other fuel grades as provided for in the legislation. Pricing in remote regions can exceed this cap. Local authorities can also cap LPG prices for residential users. Coal prices are also regulated, although they have been subject only to intermittent price controls since 2008. On-grid wholesale prices received by electricity generators and retail electricity prices are also set by the NDRC. Price caps and regulatory frameworks often differ by the type of user. Since 2008, China has introduced significant changes in its pricing regulations, both nationally and in pilot regions.

A new pricing mechanism was introduced in March 2009 for gasoline and diesel. It aimed to align gasoline and diesel prices more closely with international crude oil markets, by allowing for regular adjustments to match international fluctuations. A further refinement was made in 2013 to allow more frequent adjustment. The new mechanism resulted in sharp increases of retail prices. The difference between unregulated crude-oil prices and regulated domestic prices for refined products affects producer margins. This has led to recurring demands from (primarily state-owned) refiners for compensation by the government in the form of direct subsidies and tax or import-tariff concessions. Wholesale price ceilings prices are regulated for gasoline and diesel and ensure a guaranteed spread between wholesale and retail prices. Since retail prices are capped as well but can be undercut, private stations can engage in limited price competition. A range of other measures, including reduced import tariffs and export quota, impact prices of oil products.

Maximum prices for ethanol are set at the same rate as that for standard gasoline in each region. Since 2002, fuel ethanol producers in China have benefitted from subsidies, as well as tax reductions or exemptions of VAT or import duties, although subsidies were cut by 70% between 2009 and 2012 (Kojima, 2013). A 10% ethanol blending requirement for gasoline applies in 10 provinces. No mandates or incentives are in force for biodiesel.

Price regulations for natural gas ensure that residential users face lower prices than industrial consumers, although fertilizer producers face the lowest overall prices. The ex-plant (well-head) price is set by the central government, based on production costs plus the producer margin. At the same time, the price of natural gas imports increasing, causing a widening gap between regulated domestic gas prices and natural gas imports (IEA, 2012). As China's natural gas imports are rising, the gap between city gate prices from different sources is rising, with lower prices where gas is domestically produced, and higher prices in areas relying on imported gas. Final consumer prices result from the combination of wellhead prices with processing fees and transportation tariffs; local distribution charges (including connection fees) are regulated by provincial and local governments. Local authorities adjust their own natural-gas retail prices subject to approval from the NDRC. Prices are also regulated at different levels for households or industrial users, with the first paying lower prices, cross-subsidised by industrial users (*ibid.*).

A trial scheme for a more market-based pricing mechanism was implemented in 2011 in the Guangdong and Guangxi provinces, and then extended to 29 other provinces and municipalities (IEA, 2012). According to the scheme, the gas city-gate price is based on 85% of the weighted average of market prices of alternative fuels (LPG and fuel oil). This scheme will apply to new (incremental) gas, and the goal is to increase gradually the price of existing gas, to achieve convergence in 2015. In 2013, natural gas prices for non-residential users, which account for 80% of total gas demand, were increased by 15%. As part of the reform of natural gas pricing, the NDRC allowed local governments to establish preferential price and subsidy regimes for key users, notably in transportation (e.g., city buses and taxis) and in the residential sector (e.g., heating for low-income households), through price freezes, cash payments, or preferential tariffs. At a local level, subsidies may be provided for natural gas used by transport and for residential heating, particularly for lower-income households.

Up to the end of 2012, coal prices for non-utility use were determined by the market and coal prices for utilities were negotiated between coal producers and electricity generators within price-bands set by the NDRC (IEA, 2009). Prices for both types of use hence differed. In December 2012, the State Council abolished the different pricing arrangements and allowed the price of coal for utility use to be determined by the market. Before December 2012, electricity generators were allowed to raise electricity prices if coal prices rose by more than 5% in less than six months, which allowed them to pass through 70% of increased coal costs; since December 2012, this was modified to count increases over a 12-month period, effectively meaning they can pass through 90% of such cost increases. Several Chinese cities provide subsidies for coal-based heating, often on a temporary but recurring basis.

In the electricity sector, cost-based regulation determines the on-grid prices paid to coal-powered generators. The price is set in tandem with a quota for the amount of power each generator must supply to the grid. On average, electricity prices faced by industrial users are higher than household prices. Pricing reforms introduced in 2012 stipulate rising unit prices for households.

In practice, prices received by power generators have not kept pace with rising coal prices. High coal prices in 2011 and lower power tariffs contributed to financial losses for electric generators. When coal prices declined, the government lowered on-grid tariff rates for coal fired power plants while raising the rates for natural gas fired plants in certain regions, with the aim of encouraging electricity production from alternative fuels. Similar measures have been introduced to encourage electricity generated from renewables.

Structure of energy taxation

Introduced in January 2009, excise taxes apply to most oil products, at equal rates for all uses, including heating, industrial, agricultural use and electricity generation, although limited exemptions remain, notably for domestic aviation. Diesel, and light fuel oil are taxed at CNY 0.8 per litre; naphtha and unleaded petrol are taxed at CNY 1 per litre; and leaded petrol is taxed at CNY 1.4 per litre. These taxes are included in the graphical profiles for China.

China also administers regional-level CO_2 Emission Trading Schemes (ETS), which are in force in Beijing, Shanghai, Tianjin, Guangdong, Hubei, Chongqing and Shenzhen. All the schemes cover emissions from industry and electricity generation, four of them also cover the residential sector, and in Shanghai, the transport sector is included. The total 2013 allocations of the first six pilots (excluding Chongqing, which came into force in 2014) cover 1,115 $MtCO_2e$, making China the second largest carbon market in the world, after the EU ETS. Guangdong ETS, the largest of the Chinese pilot programs, covered 388 $MtCO_2e$ in 2013 (World Bank, 2014). The average price in the different schemes ranged from a 75.2 CNY per tonne of CO_2 in Shenzhen to 24.7 CNY per tonne of CO_2 in Hubei. Permits are auctioned only in Guangdong and Hubei; and grandfathering and free allocation of permits has applied in these and other schemes. A national ETS is tentatively scheduled to start in 2016 (Reuters, 2014). These schemes are not indicated in the graphical profiles, as they were not in force in 2012, the year for which tax information has been taken.

Value-added taxes apply at the general 17% rate to sales of crude oil, gasoline, diesel fuel, and coal for industrial use and electricity generation. Natural gas, LPG, biogas, as well as coal for residential use are, instead, subject to a preferential 13% rate; see Section 1.4.3 for more detail. Since 2008, a number of VAT exemptions and rebates have also been introduced for imports of petroleum products, coal, and natural gas. For reasons discussed in Section 2.2, these taxes are not included in the graphical profiles.

Energy use and taxation in China

As can be seen in the graphical profiles, the transport category accounts for just under 8% of energy use in China and for 6% of carbon emissions from energy use. Diesel and gasoline account for 76% of energy use in this category, in roughly equal proportions. Diesel and gasoline are both taxed under the fuel excise tax, with the rate on diesel being roughly one-third lower than that applied to gasoline in energy terms. Domestic aviation makes up just over 7% of transport energy, and is *de facto* untaxed, although it is subject to an excise tax of CNY 0.8 per litre which has been suspended since the 2009 reforms. Rail and marine fuels (just over 14% of energy and 15% carbon emissions from transport) are taxed at lower effective rates, primarily due to coal for these uses being untaxed). The remainder of the category is made up of small amounts of biofuels, LPG and natural gas.

The heating and process category accounts for the largest share of energy (53%) and carbon emissions from energy (58%). Some oil products in this category are taxed under the fuel tax, including diesel (5.6% of energy use and 4.1% of carbon emissions from energy use in this category) and fuel oil and gasoline (together 1.5% of energy and 1.1% of carbon emissions from energy from this category). Other oil products, representing 5.9% of energy and 4% of carbon emissions from this category, are untaxed. The largest source of energy in this category is coal, which generates 63.3% of heating and process energy and 68.2% of emissions, and is untaxed. Natural gas (4.9% and 2.7%, respectively) and combustibles and renewables (18.9% and 19.9%) make up the remainder of the fuels in this category and are also untaxed.

Coal is the primary source of energy used in electricity generation, accounting for 89.1% of energy use and 98.6% of carbon emissions in this category. Renewables account for the remainder, together with small amounts of natural gas and oil products. Diesel and fuel oil that are used to generate electricity are subject to the tax rates applied under the excise tax, and other generation fuels are untaxed. Electricity consumption is not taxed.

Reported tax expenditures and rebates

The "Petroleum Fuels Price-Reform Support Programmes" provides direct payments to groups most affected by the introduction of the fuel tax in 2009, including taxi drivers and users in forestry, agriculture, and public transport. Sub-national programmes that support the consumption of coal and natural gas by low-income households in specific localities are of a smaller magnitude. This is not shown separately in the graphical profiles due to the difficulty of identifying the relevant fuel use in the profiles.

Although in principle domestic aviation fuel in China is subject to a CNY 0.80 per litre excise tax, this levy has been suspended ever since the 2009 reform. This measure is represented in the graphical profiles in the subcategory "Aviation fuels (domestic)".

Naphtha used in the production of ethylene or aromatic hydrocarbons is granted a full exemption from excise taxes. From 1 January 2010, the government also exempted fuel oil from the standard excise rate under the same conditions. Those exemptions were then transformed into rebates in November 2011. These exemptions are not shown in the graphical profiles due to their small size in relation to other energy use.

In 2011, China introduced an exemption of excise tax for oil products used in geological exploration, drilling, and hydrocarbons mining. A retroactive refund of the taxes paid since 2009 was implemented, but is too small to be shown in the graphical profiles.

Key assumptions and caveats

Key assumptions and caveats are as follows:

- The category "Rail and marine fuels (domestic)" includes diesel, coal, and fuel oil for rail and marine use. Diesel and fuel oil are taxed under the fuel excise tax and coal is untaxed. The tax rate shown for this category is the average of the rates applied to diesel and fuel oil and the non-taxation of coal, weighted by the amount of energy derived from each fuel.

Information on tax rates and bases, tax expenditures energy use, and conversion rates were obtained from the sources detailed in Annex A, from Government websites, and from consultation with national officials.

In addition, country-specific sources were used as listed in the references for this chapter.

Figure 3.1. **Taxation of energy in China on an energy content basis**

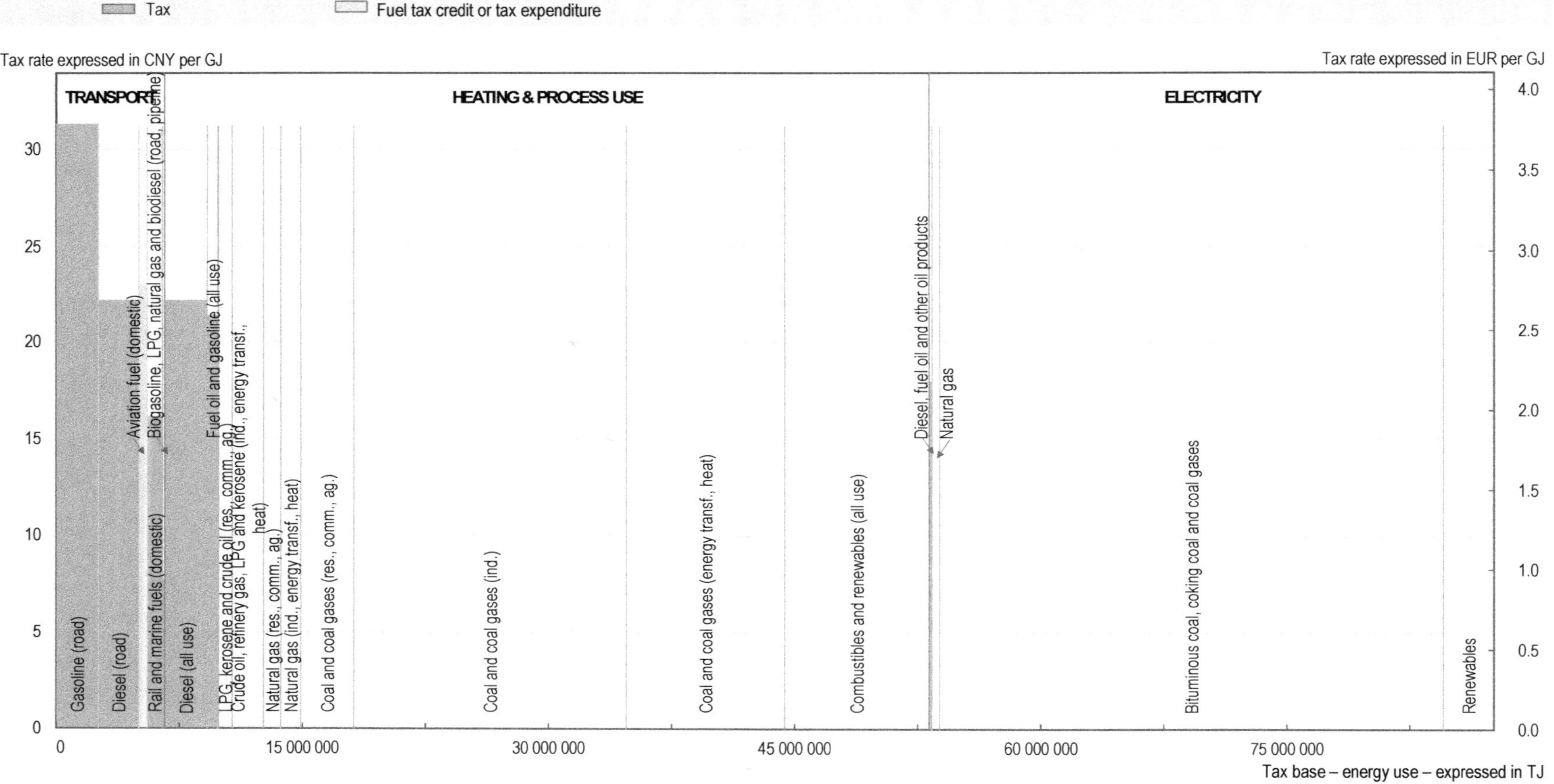

Source: OECD calculations. Tax rates are as of 1 April 2012; energy use data is for 2009 from IEA (2014b), *IEA World Energy Statistics and Balances* (database), Doi: http://dx.doi.org/10.1787/data-00513-en.

StatLink http://dx.doi.org/10.1787/888933205848

Figure 3.2. **Taxation of energy in China on a carbon content basis**

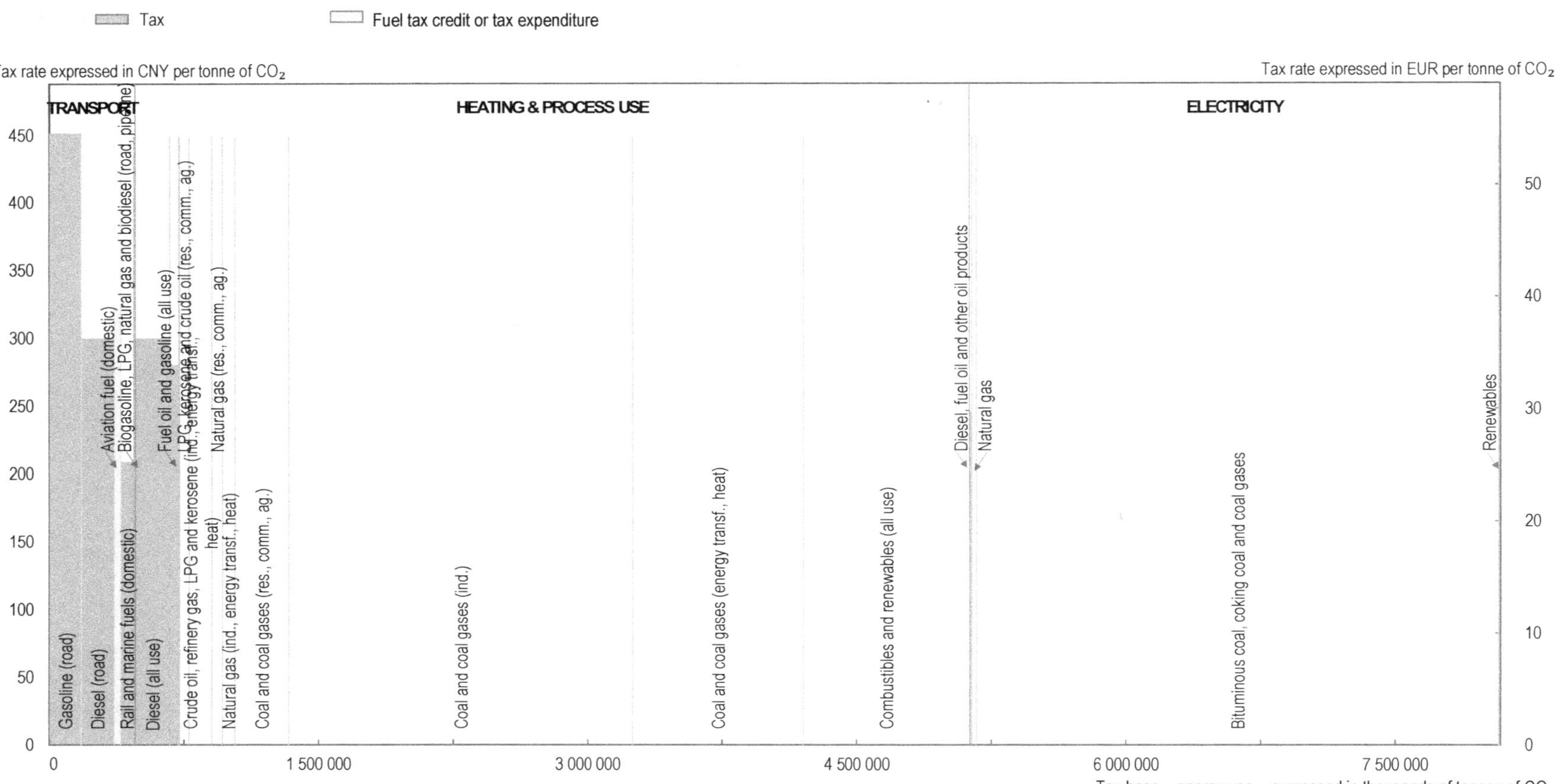

Source: OECD calculations. Tax rates are as of 1 April 2012; energy use data is for 2009 from IEA (2014b), *IEA World Energy Statistics and Balances* (database), Doi: http://dx.doi.org/10.1787/data-00513-en.

StatLink http://dx.doi.org/10.1787/888933205858

References

Chen, J. (2011), "China's Experiment on the Differential Electricity Pricing Policy and the Struggle for Energy Conservation", *Energy Policy*, 39, pp. 5076-5085.

IEA (2014a), "Gas Pricing and Regulation – China's Challenges and IEA Experience", *Partner Country Series*, *www.iea.org/publications/freepublications/publication/chinagasreport_final_web.pdf* (accessed October 2014).

IEA (2014b), "Extended World Energy Balances", *IEA World Energy Statistics and Balances* (database), Doi: *http://dx.doi.org/10.1787/data-00513-en*.

IEA (2013), "Recent Developments", OECD/IEA Publishing, Paris, *www.iea.org/media/weowebsite/2013/Developments_EnergySubsidies.pdf*.

IEA (2009), "Cleaner Coal in China", OECD/IEA Publishing, Paris.

OECD (2013), *OECD Economic Surveys – China*, OECD Publishing, Paris, Doi: *http://dx.doi.org/10.1787/eco_surveys-chn-2013-en*.

Oxford Institute for Energy Studies (2014), "The Development of Chinese Gas Pricing: Drivers, Challenges and Implications for Demand", *www.oxfordenergy.org/wpcms/wp-content/uploads/2014/07/NG-89.pdf*.

Reuters (2014), "China's National Carbon Market to Start in 2016 – Official", *http://uk.reuters.com/article/2014/08/31/china-carbontrading-idUKL3N0R107420140831* (accessed January 2015).

UNEP (2013), "China's Green Long March", *A Synthesis Report*, United Nations Environment Programme, *www.unep.org/greeneconomy/Portals/88/Research%20Products/China%20synthesis%20report_FINAL_low%20res_22nov.pdf*.

UNEP (2014), "Global Trends in Renewable Energy Investment 2014", Frankfurt School – UNEP Centre/BNEF, *www.unep.org/pdf/Green_energy_2013-Key_findings.pdf* (accessed November 2014).

Zhang, Z.X. (2014), "Energy Prices, Subsidies and Resource Tax Reform in China", *FEEM Nota di Lavoro 72/2014*, *www.feem.it/getpage.aspx?id=6599andsez=Publicationsandpadre=73*.

Zhou, N., M.D. Levine and L. Price (2010), "Overview of Current Energy-Efficiency Policies in China", *Energy Policy*, No. 38, pp. 6439-6452.

India

This chapter discusses the energy landscape, pricing policies, and the structure of energy taxation in India and presents graphical profiles of energy use and taxation in India in both energy and carbon terms (Figures 4.1 and 4.2, respectively).

India is the second most populated country in the world and the fifth largest in terms of energy consumption. Rapidly growing per-capita energy use and population growth will require immense increases in primary energy supply and electricity generation capacity. With the fifth largest coal reserves worldwide, India is the world's third largest coal producer. Domestic coal production is almost exclusively consumed domestically. However, demand is greater than domestic supply and coal imports are increasing. In 2011, India imported around 20% of coal consumption and coal imports have more than doubled from 2008 to 2013. India also imports most of its crude oil (around 70% in 2011), refining it domestically. A small amount of refined oil products is exported, particularly diesel. Around 20% of natural gas consumption is imported.

The Ministry of Petroleum and Natural Gas regulates the exploration, production, refining, distribution, marketing, imports and exports of oil and natural gas products, through two regulatory bodies (the Directorate General of Hydrocarbons, DGH, and the Petroleum and Natural Gas Regulatory Board, PNGRB). State-owned companies dominate the oil and gas sector, particularly in the refining and distribution sectors.

The Indian coal industry was nationalised in 1973 under the Coal Mines (Nationalisation) Act, which was amended in 1993 to allow coal mining by the private and public sectors for captive consumption of iron, steel and electricity production. The Ministry of Coal is responsible for the regulation of the extraction, production, supply, distribution and price of coal. It owns 90% of Coal India Limited and 100% of coal-producing subsidiaries, as well as parts or the entirety of several smaller mining operations. Until 2014, coal was allocated to different industries by the Ministry of Coal under the New Coal Distribution Policy. Coal India (74% of the market) and Singareni Collieries Company Limited had a monopoly on coal production for commercial sale. At the end of 2014, government opened its coal mining industries to private companies, removing Coal India's monopoly. About half of the coal blocks are scheduled to be auctioned to the power sector. Steel, cement and other sectors will bid for the remaining blocks (Enerdata, 2014).

India is the world's third-largest electricity producer, with coal as the primary fuel for generation. Per capita electricity consumption is significantly below the worldwide average, but it is increasing. Diversions of electricity "off the grid" result in a high proportion of electricity losses relative to electricity generation. The Ministry of Power is the central government body that regulates the electricity sector in India, with India's Central Electricity Authority (CEA) co-ordinating electricity policy. Electricity is generated by a mix of private and public generators, with private generation capacity growing faster than state-owned capacity. CEA is also responsible for overseeing the state-owned and subnationally-owned companies that generate electricity as well as PowerGrid India, a transmission company that operates around 90% of inter-State and inter-regional networks in India and transmits around 50% of India's electricity. The remainder of distribution is undertaken by a mix of government, state-owned and private companies.

India, currently home to Asia's largest solar park with a capacity of 600 MW, intends to increase the amount of clean power generation to almost MW 53 000 by 2017, almost doubling the 2012 amount of 23 128 MW (Bloomberg, 2012). This target, to be achieved using a mix of wind, solar, biomass and biofuels, as well as small hydroelectric projects, should help increase the overall electricity output and address frequent power outages (FT, 2014). Policy support is currently provided using a combination of state-level feed-in tariffs and federal subsidies (CPI, 2014).

Main energy policies influencing producer prices

Energy products are subject to a range of price controls through direct ownership, fixed retail prices, support to industry and to consumers. Subsidies provided to energy products in India are significant, estimated at around 2% of GDP (OECD, 2014). Some price reforms introduced since 2010 are aimed at cutting subsidies for some energy products.

Pricing of oil products is intended to protect domestic consumers from international price variations and ensure access to energy sources for low-income households and economic sectors considered to be a priority (e.g. agriculture). Gasoline prices were deregulated by the Central Government of India in June 2010. In late 2012, the government initiated a phased deregulation of diesel prices. However, in practice, government consent was required to raise gasoline and diesel prices, and prices were below international levels (Kojima, 2013; and Mukesh, 2012). For diesel, the resulting under-recovery per unit is roughly 16% of international prices. The Indian government has taken the oil price slump at the end of 2014 as an opportunity to remove diesel price controls, aligning diesel prices with international crude oil costs (Enerdata, 2014a), thus eliminating subsidies.

Prices for kerosene and LPG are set by the Government. The price of household kerosene has been set at INR 15 per litre since October 2012 (roughly 0.28 USD per litre) (Kojima, 2013), roughly one-third of the world price. Industrial users pay about three times this price. Subsidised kerosene and LPG cylinders (up to 12 per household) are sold through a public distribution system (PDS). These schemes result in significant under-recoveries for oil companies, and diversion of subsidised kerosene is common (International Institute for Sustainable Development, 2012). The price of diesel, PDS kerosene and LPG are seen to be of particular importance to lower-income households, and are particularly stringently monitored through regular publication of prices on government websites.

Customs duties apply to imports of a range of oil products, including gasoline, diesel, LPG, kerosene, naphtha, fuel oils, LNG and natural gas at rates ranging from 2.5-5%. Additional rates apply to many of these fuels, ranging from 8-14%, or set at a per-unit rate. Imports of crude oil are exempt from customs duty.

Coal prices were progressively deregulated between 1996 and 2000. Prices are not aligned with international prices, but are instead set at the level notified by Coal India Limited (Ministry of Coal, n.d.). Coal India sets prices for each of seven grades of coal using Useful Heat Value pricing. Import prices of coal are considerably higher than domestic prices which also entail implicit subsidies. Until 2014, the Ministry of Coal and Mines controlled the allocation of coal supplies under the New Coal Distribution Policy, prioritising power generation and fertiliser production, but the coal market was officially liberalised at the end of 2014 (Enerdata, 2014b).

Until 2014, an Empowered Group of Ministers, led by the Finance Minister, priced natural gas in consultation with the Ministry of Petroleum and Natural Gas and DGH and allocated it to domestic sectors under the Gas Utilisation Policy. Distribution of natural gas is managed by a state-owned company, GAIL. There are several categories of pricing regimes for natural gas, with power and fertiliser producers paying lower prices than other users. Prices in the Northeast of India are also lower (Deloitte, 2013). In a move to curb subsidy costs, improve competition and attract new energy investment, the Indian government raised prices for domestically-produced natural gas by 33% at the end of 2014 (Enerdata, 2014a).

In the electricity sector, consumer tariffs are set at the state level by state electricity boards. The regulated rates often differ by the sector or consumer of electricity. Electricity subsidies continue to be important, as evidenced by state electricity generation and distribution companies receiving compensation for recurring losses.

Structure of energy taxation

The Central Value Added Tax (CENVAT) is an excise duty which applies to end-user energy products. It applies to most oil products at a specific rate per litre or at an *ad valorem* rate. The rates of CENVAT in 2012 are shown in Table 4.1. An additional duty applies to gasoline and diesel at INR 2 per litre and a special additional excise duty also applies (gasoline, INR 7 per litre and diesel, INR 1 per litre).

CENVAT rates and the additional duties on excise taxes are shown in the graphical profiles, subject to the assumptions described at the end of this chapter.

Table 4.1. **2012 rates of excise duty on energy products**

Product		Basic cenvat duty	Special additional excise duty	Additional excise duty
Crude petroleum		Nil + INR 4 500/MT + INR 50/MT as NCCD		
Petrol		INR 1.20/litre	INR 6/litre	INR 2.00/litre
Petrol (branded)		INR 7.50/litre	INR 6/litre	INR 2.00/litre
High speed diesel		INR 1.46/litre	Nil	INR 2.00/litre
High speed diesel (branded)		INR 3.75/litre	Nil	INR 2.00/litre
LPG	Domestic	Nil	Nil	Nil
	Non-domestic	8.0%	Nil	Nil
Kerosene	PDS	Nil	Nil	Nil
	Non PDS	14.0%	Nil	Nil
Aviation turbine fuel		8%	Nil	Nil
Naphtha	Non-fertilizer	14.0%	Nil	Nil
	Fertilizer	Nil	Nil	Nil
Bitumen and asphalt		14.0%	Nil	Nil
Furnace oil	Fertilizer	Nil	Nil	Nil
	Non-fertilizer	14.0%	Nil	Nil
Light diesel oil		14% + INR 2.50/litre	Nil	Nil
Liquefied natural gas		Nil	Nil	Nil
Low sulphur heavy stock/ HPS and other res.	Fertilizer	Nil	Nil	Nil
	Non-fertilizer	14.0%	Nil	Nil
Lube oil/greases		14.0%	Nil	Nil
Natural gas (gaseous state)		Nil	Nil	Nil
Petroleum coke		14.0%	Nil	Nil

Source: Petroleum Planning and Analysis Cell (2012)

A Clean Energy Cess applies to coal. This tax, which was introduced in July 2010, applied at a rate of INR 50 per tonne to all coal, whether imported or domestic, including coal used in electricity generation. The tax was to be a first step towards a carbon tax and revenue from the tax (estimated at around INR 25-30 billion in 2010) is earmarked for the National Clean Energy Fund (NCEF). The NCEF funds research and innovation in clean energy technology. The rate of the Clean Energy Cess was increased to INR 100 per tonne in July 2014 and to INR 200 in February 2015. The Clean Energy Cess is shown in the graphical profiles at the rate of INR 50 per tonne that applied at 1 April 2012.

An Education Cess applies to other taxes, including CENVAT and the Clean Energy Cess, at a rate of 3% of the rate of the tax (see for example Indian Oil, 2014a).

Oil and coal used in the generation of electricity are subject to CENVAT and to the Clean Energy Cess. In addition, the consumption of electricity is also taxed, with responsibility for the structure and level of taxation lying at state level. Tax rates on the consumption of electricity therefore vary between states, both in the way rates are set (either by reference to the amount or value of electricity consumed) and in differentiations based on use. Table 4.2 sets out rates in some states, based on the OECD database on environmentally-related taxes. The graphical profiles indicate the tax rates applied to electricity in Gujarat and Tamil Nadu (the states that consumed the most and third-most electricity in 2009: Energy Management Group, 2009).

Table 4.2. **Rates on electricity consumption in some Indian states**

State	Use	Rate (INR/KWh)	Rate (% of price)
Assam	Industrial use	0.01-0.03	
	Non-industrial use	0.05	
Bihar	Agricultural, irrigation and industrial purposes	0.02	
	Domestic	0.08	
	Mining: < 100 horsepower	0.15	
	> 100 horsepower	0.08	
	Other commercial purposes	0.12	
Gujarat		0.02	
Karnataka			5%
Orissa			15%
Rajasthan	Commercial and domestic	0.06	
	Agriculture: metered	0.01	
	not metered		5%
	Industry and mining	0.01	
Tamil Nadu		0.1	

Source: OECD Database of Environmentally-Related Taxes (2014).

At subnational level, states apply sales taxes or value-added taxes across a wide range of intermediate and final goods, including energy taxes, with rates that vary by state. Reduced tax rates apply to coal, crude oil, aviation fuel, and LPG for domestic use, which are "declared goods", prohibiting states from taxing them at a rate higher than 4% (normal rates range from 5-33%; OECD, 2014). See Section 1.4.3 for details on VAT policy. For the reasons discussed in Section 2.2, these taxes are not included in the graphical profiles.

Energy use and taxation

As can be seen in the graphical profiles, the transport category accounts for just over 8% of energy use in India and for 6% of carbon emissions from energy use. It is dominated by diesel and gasoline, accounting for 90% of energy and emissions used in this category, over two-thirds of which are from diesel. Both fuels are taxed; diesel is taxed at just over one-third of the rate applied to gasoline. The remainder of the transport category is made up of aviation and marine fuels (just over 5.5% of energy and carbon emissions from transport); and small amounts of natural gas, fuel oil and naphtha, which are taxed at lower rates.

The heating and process category accounts for 52% of energy and 56% of carbon emissions from energy. Within this category, oil products are taxed at the highest rates, with fuel oil and naphtha (2.9% of heating and process energy, and 2.2% of heating and process carbon emissions) being the highest taxed. Diesel, representing 5% of energy and 3.76% of emissions in this category, is taxed at lower rates than fuel oil and naphtha. Energy from coal represents 25.1% of heating and process energy and 26.1% of carbon emissions from this category, and is taxed under the Clean Energy Cess, although at lower effective rates than fuel oil, naphtha and gasoline. The remainder of oil products in this category (7.4% of energy and 5% of carbon emission from this category), natural gas (4% and 2.3%, respectively) and biomass (51.4% and 58.3%) are untaxed.

Coal is the primary source of energy used in electricity generation, accounting for 80.4% of energy use and 89.6% of carbon emissions in this category. Natural gas, oil and renewables account for the remainder, together with a small amount of solid biomass. Oil products and coal that are used to generate electricity are subject to the tax rates applied in the other categories. These taxes are shown in the graphical profiles. In addition, states levy taxes on electricity consumption. The graphical profiles also indicate the tax rate on electricity consumption in Gujarat and Tamil Nadu, using the "look-through" approach described in the methodology section.

Reported tax expenditures and rebates from taxes shown in the graphical profiles

Sales of domestically-produced crude oil are exempt from excise taxes, to encourage investment in the upstream oil sector by both domestic and international companies (OECD, 2014). This is not shown in the graphical profiles for India as the exemption applies to crude rather than refined oil products for final consumption and therefore cannot be separately identified.

The government reports the level of subsidy provided through the fixed per-unit grants for PDS kerosene and LPG cylinders (OECD, 2014). This tax expenditure is not shown in the graphical profiles due to the difficulty in estimating the amount of kerosene and LPG to which this measure applies.

Key assumptions and caveats

Key assumptions and caveats are as follows:

- All kerosene and LPG have been assumed to be exempt from CENVAT. This assumption is based on a comparison of the amounts of both fuels used under the IEA data and data on fuel volume obtained from Petroleum Planning and Analysis Cell (2014).

- Rates indicated in the graphical profiles are those for Gujarat and Tamil Nadu (the states that consumed the most and third-most electricity in 2009) and have been taken from the OECD database on environmentally-related taxes.
- Tax rates on aviation fuels, fuel oil and naphtha are set in *ad valorem* terms. Information on fuel prices in 2012 for these fuels was not found. Instead the *ad valorem* rates have been converted into per-unit rates based on data on aviation fuel and crude oil prices from Indian Oil (2014b).

Information on tax rates and bases, tax expenditures energy use, and conversion rates were obtained from the sources detailed in Annex A, and from Government websites.

In addition, country-specific sources were used as listed in the references chapter.

Figure 4.1. **Taxation of energy in India on an energy content basis**

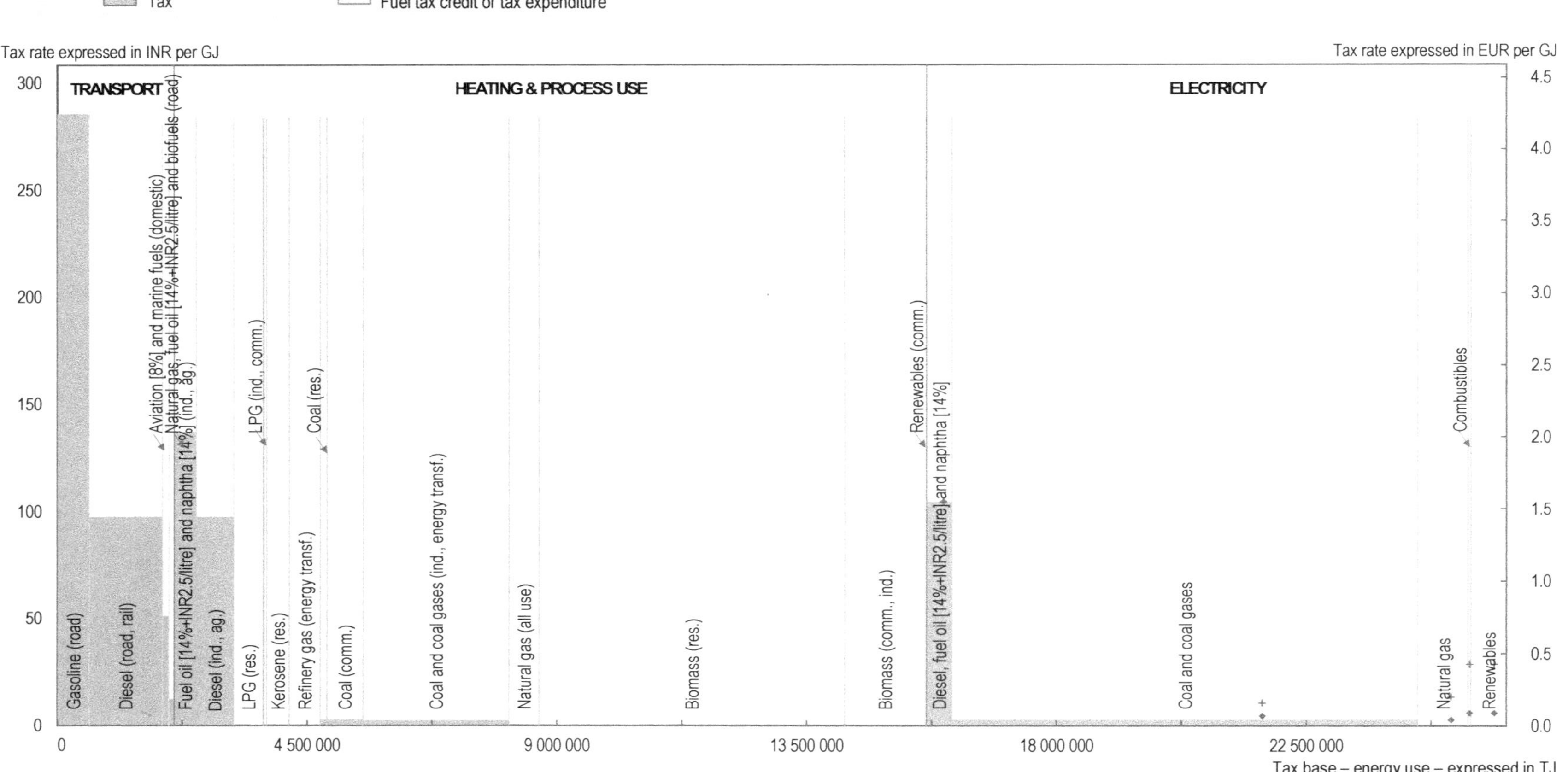

Source: OECD calculations. Tax rates are as of 1 April 2012; energy use data is for 2009 from IEA (2014), *IEA World Energy Statistics and Balances* (database), Doi: http://dx.doi.org/10.1787/data-00513-en. Electricity consumption tax rates shown include federal level taxes on inputs to electricity generation; taxes on the consumption of electricity are indicated for Gujarat and Tamil Nadu. Tax rates on aviation fuel, fuel oil and naphtha have been calculated as a percentage of their price, subject to the caveats described in this chapter.

StatLink http://dx.doi.org/10.1787/888933205869

Figure 4.2. **Taxation of energy in India on a carbon content basis**

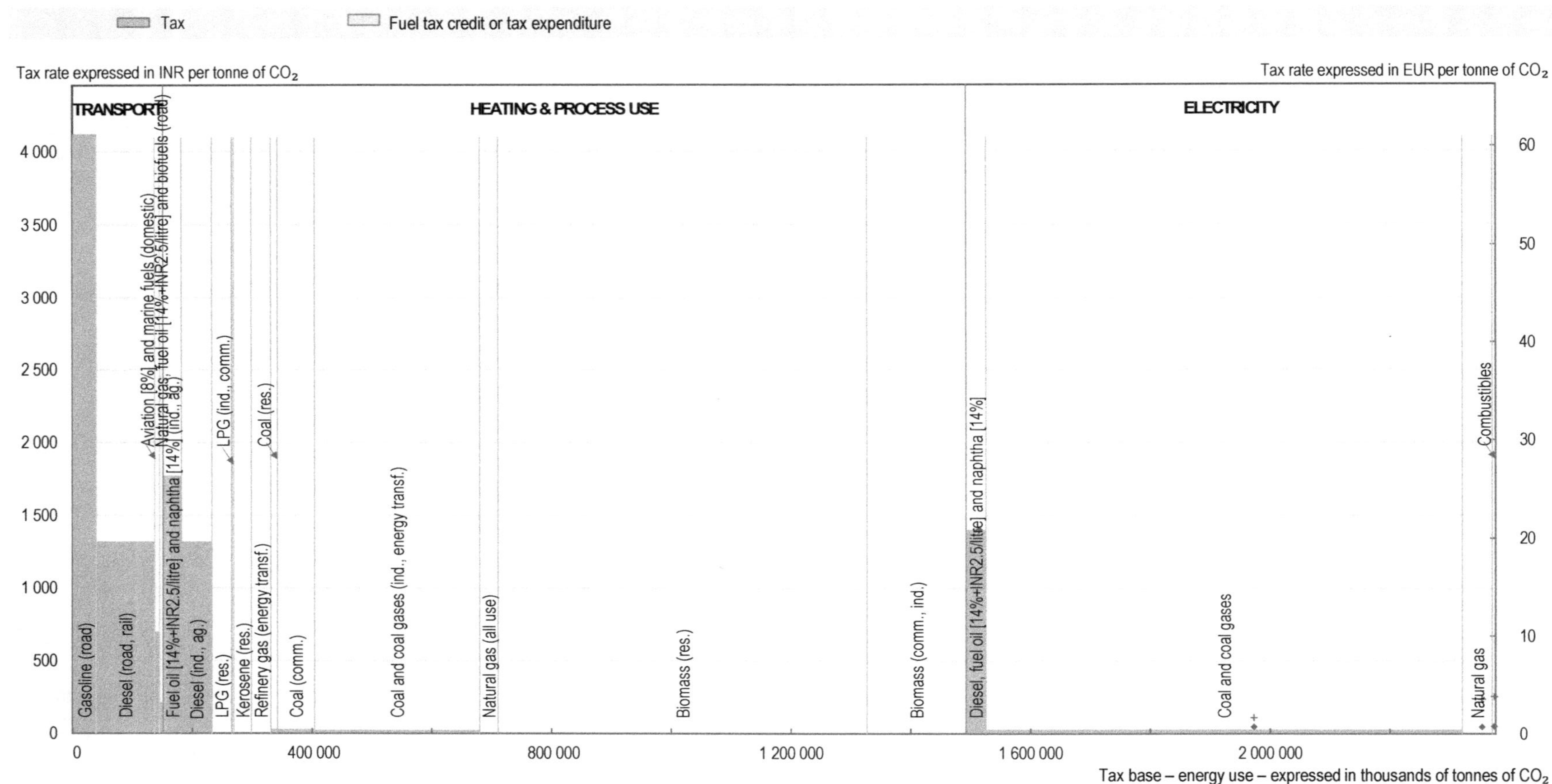

Source: OECD calculations. Tax rates are as of 1 April 2012; energy use data is for 2009 from IEA (2014), *IEA World Energy Statistics and Balances* (database), Doi: http://dx.doi.org/10.1787/data-00513-en. Electricity consumption tax rates shown include federal level taxes on inputs to electricity generation; taxes on the consumption of electricity are indicated for Gujarat and Tamil Nadu. Tax rates on aviation fuel, fuel oil and naphtha have been calculated as a percentage of their price, subject to the caveats described in this chapter.

StatLink http://dx.doi.org/10.1787/888933205870

References

Balachandran, S. (2008), "Cenvat and its Implications", Welingkar Institute of Management Development and Research, *www.welingkaronline.org/DLP/FINANCE%20-%20CENVAT%20AND%20 ITS%20IMPLICATIONS.pdf.*

Bloomberg (2012) , "India Targets Doubling of Renewable Energy Installations to 2017", *www.bloomberg.com/news/2012-05-22/india-targets-doubling-of-renewable-energy-installations-to-2017.html* (accessed October 2014).

Central Excise and Service Tax Commissioner Patna (n.d.), "Central Excise Guide", *http://cexpatna.bih.nic.in/CENTRAL%20EXCISE%20GUIDE.pdf* (accessed October 2014).

Climate Policy Initiative (2014), "Solving India's Renewable Energy Financing Challenge: Which Federal Policies Can Be Most Effective?", *http://climatepolicyinitiative.org/wp-content/uploads/2014/03/Which-Federal-Policies-can-be-Most-Effective-Full-Study.pdf.*

Corbeay, A.S. (2010), "Natural Gas in India", *IEA Working Paper*, Paris, *www.iea.org/publications/freepublications/publication/natural_gas_india_2010.pdf.*

Deloitte (2014), "Gas Pricing in India", *Deloitte Perspectives*, *http://www2.deloitte.com/in/en/pages/energy-and-resources/articles/gas-pricing-in-india.html* (accessed October 2014).

Enerdata (2014a), "India Liberalises Diesel Price and Raises Gas Price by 33%", *Enerdata Recent Energy News*, *www.enerdata.net/enerdatauk/press-and-publication/energy-news-001/india-liberalises-diesel-pri ce-and-raises-gas-price-33_30366.html* (accessed October 2014).

Enerdata (2014b), "India Will Auction or Allot 18 More Coal Blocks", *Enerdata Recent Energy News*, *www.enerdata.net/enerdatauk/press-and-publication/energy-news-001/india-will-auction-or-allot-18-mo re-coal-blocks_30911.html* (accessed December 2014).

Energy Management Group (2009), "State-Wise Electricity Consumption and Conservation Potential in India", *www.emt-india.net/eca2009/14Dec2009/CombinedSummaryReport.pdf.*

Ernst and Young (2014), "Natural Gas Pricing in India", *www.ey.com/Publication/vwLUAssets/EY-natural-gas-pricing-in-India/$FILE/EY-natural-gas-pricing-in-India.pdf.*

Financial Times (2014), "India Targets Renewables in $250bn Power Plan", *www.ft.com/cms/s/0/36b37c48-657e-11e4-aba7-00144feabdc0.html#axzz3LJsipOOJ.*

IEA (2014), "Extended World Energy Balances", *IEA World Energy Statistics and Balances* (database), Doi: *http://dx.doi.org/10.1787/data-00513-en.*

Indian Oil (2014a), "Price Build-Up of Petrol at Delhi Effective 16/07/2014", *https://iocl.com/Products/PriceBuildup/Price_buildup_of_MS.pdf.*

Indian Oil (2014b), "Products", *http://www.iocl.com/products.aspx.*

International Institute for Sustainable Development (2012a), "India's Fuel Subsidies: Policy Recommendations for Reform", *IISD Policy Brief*, *www.iisd.org/gsi/sites/default/files/ffs_india_guide_rev.pdf.*

International Institute for Sustainable Development (2012b), "Fossil-Fuel Subsidy Reform in India: Cash Transfers for PDS Kerosene and Domestic LPG", *www.iisd.org/gsi/sites/default/files/ffs_india_teri_rev.pdf.*

Kanani, H. (2011), "Pricing of Natural Gas in India", *www.greatlakes.edu.in/gurgaon/sites/default/files/Pricing_of_Natural_Gas_in_India.pdf.*

KPMG, (2013), "Taxes and Incentives for Renewable Energy", *www.kpmg.com/global/en/issuesandinsights/articlespublications/taxes-and-incentives-for-renewable-energy/pages/india.aspx.*

Ministry of Coal, (n.d.), "Pricing of coal", *www.coal.nic.in/policy/pricing.pdf.*

Ministry of Petroleum and Natural Gas (28/01/2011), "PDS Kerosene and Domestic LPG Subsidy Scheme, 2002, *http://ppac.org.in/PDF/Govt_Link8.pdf.*

Mukesh, A. (2012), "Diesel Pricing in India: Entangled in a Policy Maze", *National Institute of Public Finance and Policy*, *www.nipfp.org.in/media/medialibrary/2013/08/Diesel_Price_Reform.pdf.*

OECD (2014), *OECD Economic Surveys: India 2014*, OECD Publishing, Paris, Doi: *http://dx.doi.org/10.1787/eco_surveys-ind-2014-en.*

OECD (2011), *OECD Economic Surveys: India 2011*, OECD Publishing, Paris, Doi: *http://dx.doi.org/10.1787/eco_surveys-ind-2011-en.*

Petroleum Planning and Analysis Cell, (2014), "Historical Consumption of Petroleum Products", *http://ppac.org.in/content/147_1_ConsumptionPetroleum.aspx.*

Petroleum Planning and Analysis Cell, (2012), "Central Excise and Customs Tariff Table", *http://ppac.org.in/* (accessed October 2014).

Petroleum Planning and Analysis Cell (n.d.), "Price Build-Up of Diesel, PDS Kerosene and Domestic LPG", *http://ppac.org.in/writereaddata/Price%20Build%20Up%20for%20Sensitive%20Products.pdf.*

Indonesia

This chapter discusses the energy landscape, pricing policies, and the structure of energy taxation in Indonesia and presents graphical profiles of energy use and taxation in Indonesia in both energy and carbon terms (Figures 5.1 and 5.2, respectively).

Indonesia is the world's leading coal exporter. In the past decade, Indonesia's coal consumption nearly tripled. The government intends to increase coal consumption to cover one-third of domestic energy use by 2025, compared to a 15% share in 2009 (IEA, 2011; OECD, 2014). The country was the world's 24th largest crude oil producer in 2013, and became a net oil importer in 2004 (EIA, 2014). It left OPEC in 2009 to use its oil production to meet growing domestic demand. Natural gas production increased by almost 25% between 2002 and 2012 and exports are declining in absolute terms, but still amount to about half of total production. Indonesia is Asia's largest biodiesel producer, and in 2012 about a third of production was used domestically.

The Ministry of Energy and Mineral Resources (MEMR) has primary responsibility for energy policy. The 2001 Oil and Gas Law significantly restructured the oil and gas sectors, creating two state-owned entities, BPMigas and BPHMigas, which were assigned regulatory responsibility for these sectors. Since the Constitutional Court ruled BPMigas unconstitutional in November 2012, a temporary task force is responsible for production sharing contracts, determining sellers of government oil and gas and managing oil and gas production. Oil and natural gas exploration and production are dominated by international oil companies, among which Chevron is the largest, accounting for 39% of the country's crude production in 2013 (EIA, 2014). The state-owned oil company Pertamina accounts for 17% of oil production and, through a subsidiary, for 13% of natural gas production. Pertamina operates nearly all refinery capacity, it controls crude oil and oil products imports, and supplies subsidised fuels (*ibid.*).

Coal production is in private hands and is relatively concentrated. In 2011, the top six producers accounted for three-quarters of total production. Among these, PT Adaro Energy and PT Bumi Resources and their subsidiaries are among the largest. To stimulate domestic and foreign investment in the sector, the 2009 mining law intends to stimulate the domestic mineral processing industry, increase transparency and standardise tenders and licenses for coal mining blocks.

Despite longstanding plans to increase competition and eliminate the state-monopoly for electricity distribution, the electricity sector is still under the control of the state-owned electricity company Perusahaan Listrik Negara (PLN). The company owns and operates 85% of generation capacity and maintains an effective monopoly over distribution and retail activities. A plan to stimulate foreign investment includes mandating guaranteed power purchase agreements for independent producers.

Indonesia has a rich endowment of renewable energy resources, such as hydropower and geothermal, whose potential is largely untapped (IISD, 2012). A 2006 presidential decree set out to increase the share of renewable energies in the total energy mix to 17% by 2025, of which biofuels and geothermal sources are supposed to cover 5%-point each, the remainder being filled by solar, hydro and wind power. The target was revised to 25% in 2010. Since then, several policies were introduced to support renewable energy development, including a feed-in tariff for renewable electricity (*ibid.*).

Main energy policies affecting producer prices

A quarter of crude oil production must be sold domestically, at rates set by BPMGas that are 25% below international prices (Domestic Market Obligation, DMO). Though recent years have seen substantial progress in abolishing subsidies, limiting their strain on government budgets, Indonesia still has one of the largest fossil fuel subsidy programmes in the region, causing fuel prices to be among the cheapest in the world (OECD, 2014). In 2012, subsidies amounted to about 15% of general government expenditures and 60% of public expenditures on education and health (*ibid.*).While retail subsidies for oil products for industrial users and for high-grade transport fuels were phased out in 2005, prices of kerosene for non-industrial users and LPG for households and businesses continue to be government-regulated at below-market rates. In 2008, gasoline was around 60% cheaper in Indonesia than in neighbouring countries and in 2012, diesel prices were 35% lower than international prices (IEA and ERIA, 2013; Asian Development Bank, 2014).

Transport fuel prices have been repeatedly increased, in line with the government's commitment to fossil fuel subsidy reform, accompanied by cash transfers for low-income households and additional programs to limit the distributional consequences of price increases. Price changes are implemented on an *ad hoc* basis, usually communicated by MEMR via press releases (IISD, 2010). In 2012, new policies allowed for price increases if the six-month average price of Indonesian crude oil exceeded a specified threshold. In mid-2013 and at the end of 2014, subsidised gasoline and diesel prices were raised by more than 30% each combined with social assistance to more than 15 million people (Reuters, 2014). Reforms at the beginning of 2015 removed subsidies for low-octane gasoline and fixed diesel subsidies at USD 0.08 per litre (OECD, 2015).

Fuel subsidies are determined based on the difference between a national reference price set by the government for each of these fuels and international benchmark prices, usually Mids Oil Platt Singapore. Subsidies are paid to Pertamina, which is reimbursed every three months for its losses from selling fuel at below-market prices. The government is also discussing the idea of moving away from a subsidy that sets the retail price of fuel to a fixed level of subsidy, thereby linking domestic prices to world prices.

In the effort to abolish kerosene subsidies, households and small businesses receive subsidised LPG and LPG-fired stoves via the "kerosene-to-LPG conversion program" since 2007, while all other users buy LPG at market prices. The programme has substantially reduced kerosene consumption. Companies are obliged to sell at least a quarter of natural gas production output on the domestic market at far below international prices, and all remaining natural gas output is sold at market prices.

Coal prices are regulated by MEMR via the Indonesian Coal Price Reference according to a specific formula, and prices are well below international prices. Companies are typically obliged to sell about a quarter of their production on the domestic market, but domestic market obligations are adjusted annually. The prime beneficiary of this policy is the state-monopoly in the power sector, PLN, buying around 70% of subsidised coal sold. To account for lower demand, the 24% DMO was temporarily reduced to 20% in October 2012.

Electricity prices are regulated at below cost levels by MEMR for all users. The government compensates PLN for the difference between their costs of electricity generation and retail prices. Faced with rising prices for fuel inputs, the government started to increase electricity tariffs in 2010, but met with strong political resistance which led to a postponement of the scheduled price increases. In 2013, prices were increased by 15% for all users, except for those with very low consumption, but continue to be far below the cost-recovery level.

Biofuel subsidies have been in force since 2009. In 2013, biodiesel was subsidised by IDR 3 000 per litre and ethanol by IDR 3 500 per litre. Ethanol fuel production based on molasses was suspended in 2010 because of high production costs and feedstock prices despite a subsidy of IDR 2 000 per litre (IISD, 2012).

Structure of energy taxation

Indonesia does not levy any excise taxes on energy products at the national level. The Automotive Fuel Tax is a regional tax imposed on transport fuel sales (OECD 2013; IISD, n.d.). Although the government recommends a maximum tax rate of 10% of the sales price, a 5% rate was applied until September 2012. The tax rates applied under the Automotive Fuel Tax may differ by province (Switch Asia, 2013). This tax is shown in the graphical profiles.

The VAT treatment of energy use in Indonesia is summarised in section 1.4.3. While the standard VAT rate is at 10% of the sales price, oil, natural gas, coal and geothermal energy are VAT exempt, as is electricity for households using less than 6.6 kW. Similarly, as part of the kerosene-to-LPG programme, 3-kg LPG cylinders are VAT exempt and VAT on domestic biofuels sales is borne by the government. For the reasons discussed in section 2.2, VAT is not included in the graphical profiles.

Energy use and taxation in Indonesia

As shown in the graphical profiles, transport accounts for just under 17% of total energy use and 15% of total carbon emissions. Oil products account for almost all consumption of the transport sector, with gasoline representing 59% and diesel 35%. Gasoline and diesel are taxed under the Automotive Fuel Tax, and together account for 90% of energy use and carbon emissions in transport. The remaining fuels used in transport remain untaxed.

Energy for heating and process use accounts for 56% of energy use and 65% of carbon emissions. Within this category, half of the energy is from combustibles and waste, mostly consumed by households. Combustibles and waste account for 40% of emissions from this category. Oil products and natural gas account for around 20% of energy use each, but oil products account for 16% of carbon emissions, while carbon emissions from natural gas account for 12%. Energy use from coal is almost 10% of the total, and accounts for just over 10% of carbon emissions. All coal is consumed for industrial and energy transformation purposes. Energy for heating and process use is entirely untaxed.

The remaining 27% of energy is used to generate electricity, which accounts for just under 20% of total emissions. About 34% of electricity is generated from coal. Electricity generation from coal causes 60% of the carbon emissions from that sector. Natural gas and oil products each account for about 16% of electricity generation, whereas natural gas accounts for 17% of carbon emissions from electricity generation and oil products for

23%. The remaining 33% of electricity is generated from biomass, geothermal and hydro sources, causing a negligible share of emissions. Fuel inputs to electricity generation are untaxed, as is the consumption of electricity.

Reported tax expenditures and rebates relevant for graphical profiles

Indonesia does not report any tax expenditures or rebates relevant to the graphical profiles.

Key assumptions and caveats

Tax levels for the Automotive Fuel Tax were computed using 2012 retail price data for high-grade diesel ("Dex") and gasoline ("Pertamina Plus") provided by Pertamina (MEMR, 2012) for the period from April 2011 to March 2012, which is the most recent data available. It was assumed that a 5% rate applied across all states.

Information on tax rates and bases, tax expenditures energy use, and conversion rates were obtained from the sources detailed in Annex A and from Government websites.

In addition, country-specific sources were used as listed in the references for this chapter.

Figure 5.1. **Taxation of energy in Indonesia on an energy content basis**

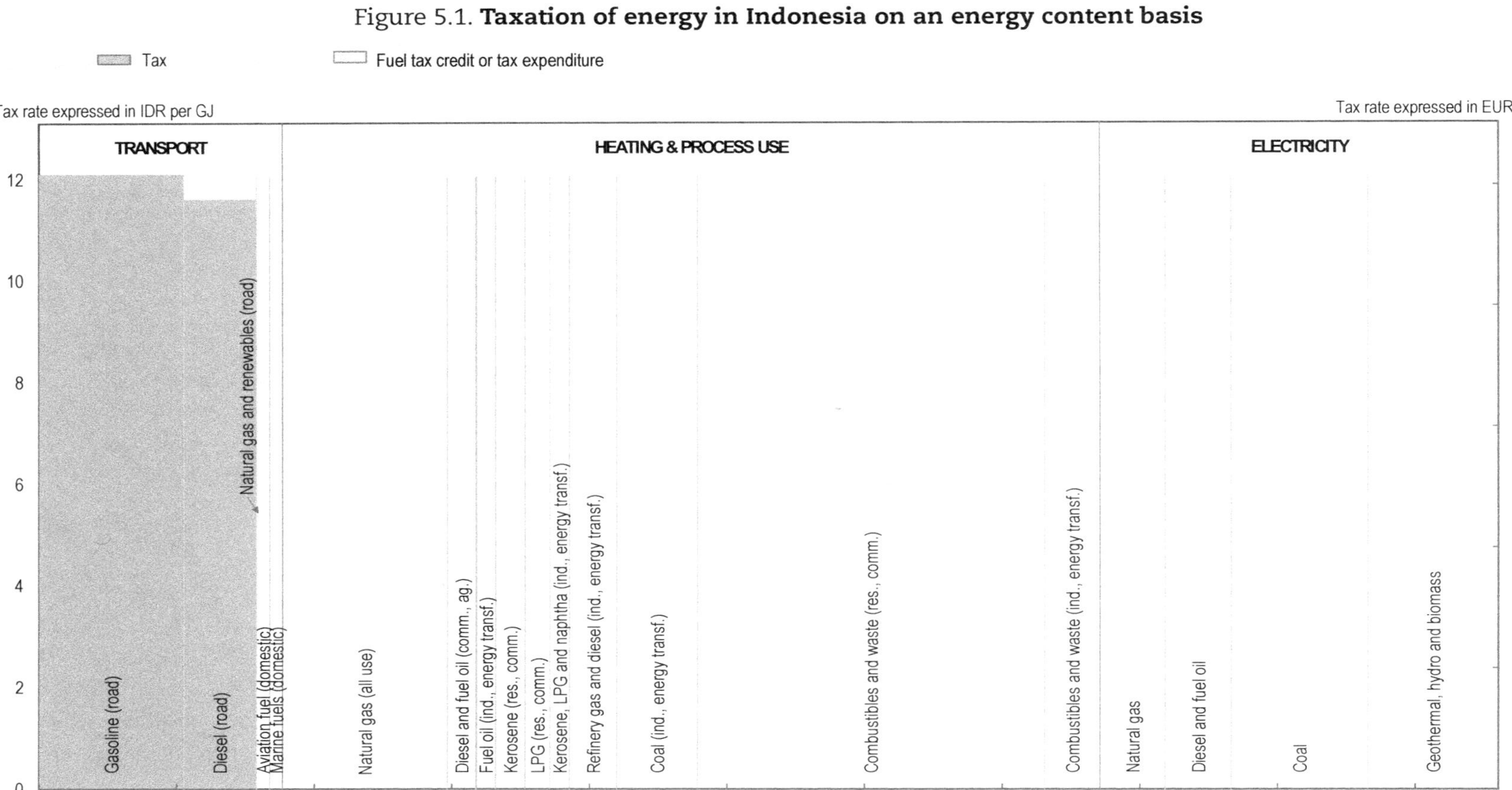

Source: OECD calculations. Tax rates are as of 1 April 2012; energy use data is for 2009 from IEA (2014), *IEA World Energy Statistics and Balances* (database), Doi: http://dx.doi.org/10.1787/data-00513-en. Tax rates on gasoline (road) and diesel (road) have been calculated as a percentage of their price, subject to the caveats described in this chapter.

StatLink http://dx.doi.org/10.1787/888933205889

Figure 5.2. **Taxation of energy in Indonesia on a carbon content basis**

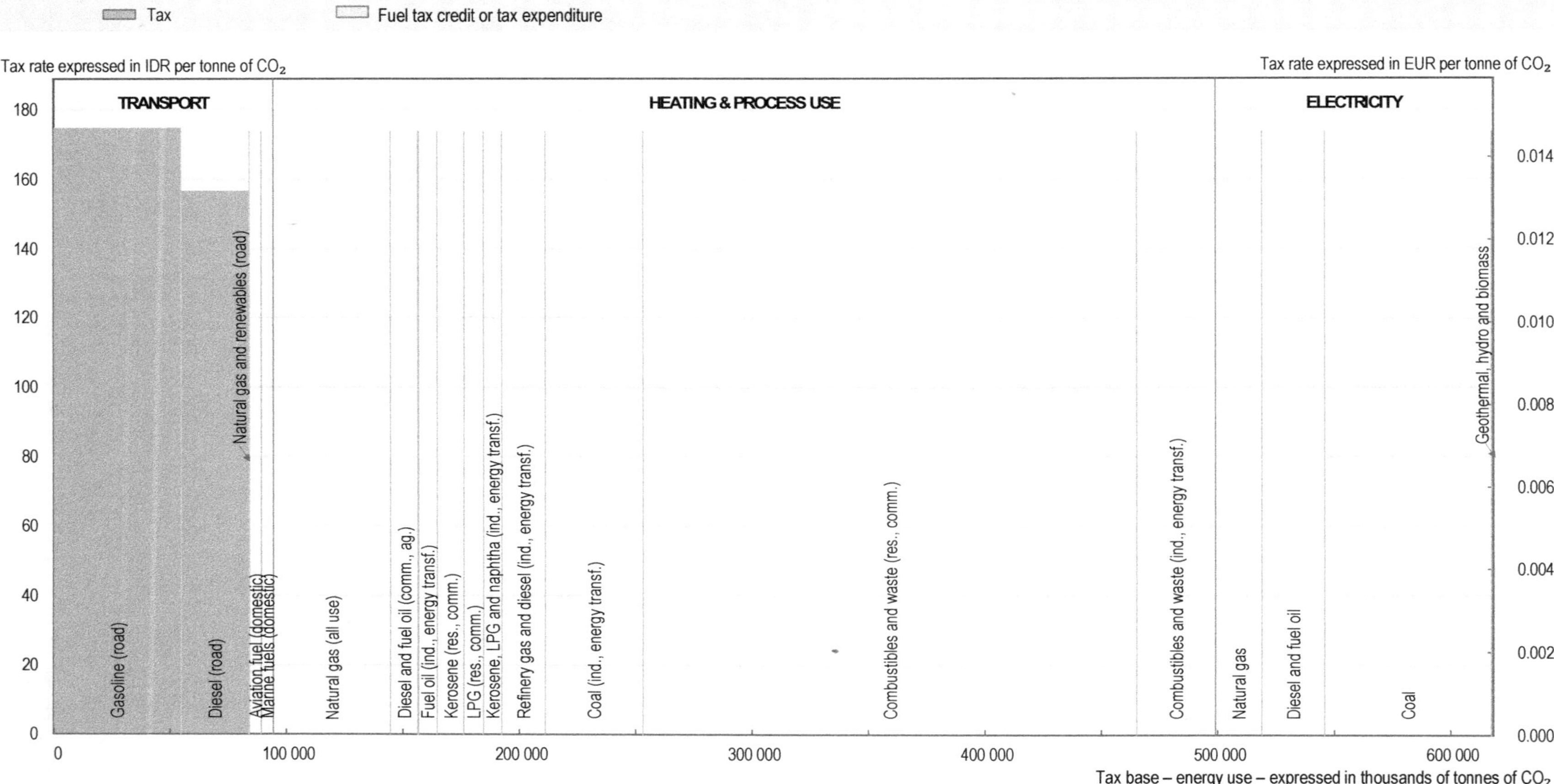

Source: OECD calculations. Tax rates are as of 1 April 2012; energy use data is for 2009 from IEA (2014), *IEA World Energy Statistics and Balances* (database), Doi: http://dx.doi.org/10.1787/data-00513-en. Tax rates on gasoline (road) and diesel (road) have been calculated as a percentage of their price, subject to the caveats described in this chapter.

StatLink *http://dx.doi.org/10.1787/888933205891*

References

Asian Development Bank (2014), "Reta 7834: Assessment and Implications of Rationalizing and Phasing out Fossil Fuel Subsidies", *Finalization Conference Report, www.adb.org/publications/reta-7834-finalization-conference-report*.

Fiscal Policy Office, (n.d.), "Fuel Subsidy Policy in Indonesia", *Presentation held at IISD Conference "Increasing the Momentum of Fossil-Fuel Subsidy Reform: Developments and Opportunities, www.iisd.org/gsi/sites/default/files/ffs_gsiunepconf_sess2_askolani.pdf.*

Global Subsidies Initiative (2013), Indonesia Energy Subsidy Update: January 2013, *www.iisd.org/gsi/news/indonesia-energy-subsidy-update-january-2013.*

IEA (2008), "Indonesia Energy Policy Review", *OECD/IEA Publishing*, Paris, *www.iea.org/publications/freepublications/publication/name-15765-en.html.*

IEA (2014), "Extended World Energy Balances", *IEA World Energy Statistics and Balances* (database), Doi: *http://dx.doi.org/10.1787/data-00513-en.*

International Institute for Sustainable Development (2013a), "Investment Incentives for Renewable Energy: Case Study of Indonesia", *TKN Report, www.iisd.org/tkn/pdf/investment_incentives_indonesia.pdf.*

International Energy Agency (2013b), "Southeast Asia Energy Outlook – World Energy Outlook 2013 Special Report", *OECD/IEA Publishing*, Paris, *www.iea.org/publications/freepublications/publication/SoutheastAsiaEnergyOutlook_WEO2013SpecialReport.pdf.*

International Institute for Sustainable Development (2010), "Lessons Learned from Indonesia's Attempts to Reform Fossil-Fuel Subsidies", *Trade Investment and Climate Change Series, www.iisd.org/publications/lessons-learned-indonesias-attempts-reform-fossil-fuel-subsidies.*

IMF (2013), "Case Studies on Energy Subsidy Reform: Lessons and Implications", *International Monetary Fund, Washington, www.imf.org/external/np/pp/eng/2013/012813a.pdf.*

Migas (2010), "Penandatanganan KKS WK Migas Lelang Reguler 2010", *www.migas.esdm.go.id/berita-kemigasan/detail/2037/Indonesia-to-Have-71-Gas-Filling-Stations-and-13-MRU-in-2014* (accessed October 2014).

Ministry of Energy and Mineral Resources of Republic of Indonesia (2012), "Daftar Harga Pertamax, Pertamax Plus Dan Pertamina Dex Tahun 2012", *www.esdm.go.id/publikasi/harga-energi/harga-bbm-dalam-negeri.html* (accessed October 2014).

OECD (2015), *OECD Economic Surveys: Indonesia 2015*, OECD Publishing, Paris, Doi: *http://dx.doi.org/10.1787/eco_surveys-idn-2015-en.*

OECD (2014), *Towards Green Growth in Southeast Asia*, OECD Publishing, Paris, Doi: *http://dx.doi.org/10.1787/9789264224100-en.*

OECD (2012), *OECD Economic Surveys: Indonesia 2012*, OECD Publishing, Paris, Doi: *http://dx.doi.org/10.1787/eco_surveys-idn-2012-en.*

Reuters (2014), "Indonesia Hikes Fuel Prices, Saving Government $ 8 billion Next Year", *www.reuters.com/article/2014/11/17/us-indonesia-subsidies-hike-idUSKCN0J11KN20141117.*

Switch Asia (2013), "The Sustainable Consumption and Production (SCP) Policy Project – Indonesia – Incentives and Policy Instruments for Promotion of Sustainable Production", *www.switch-asia.eu/fileadmin/user_upload/Indonesia_2013DEC09_SCP_Incentives_and_instruments_final_report.pdf.*

World Bank (2012), "Indonesia Economic Quarterly Report, April 2012", *http://documents.worldbank.org/curated/en/2012/04/16207744/indonesia-economic-quarterly-redirecting-spending.*

Russian Federation

This chapter discusses the energy landscape, pricing policies, and the structure of energy taxation in Russia and presents graphical profiles of energy use and taxation in Russia in both energy and carbon terms (Figures 6.1 and 6.2, respectively).

Russia is the world's third largest oil producer and it holds the world's largest reserves of natural gas. The country's economy largely depends on oil, natural gas and coal exports.

With the exception of nuclear power (controlled by Rosatom), the Ministry of Energy and the Ministry of Economic Development have primary responsibility for energy policy. Regulatory power in fossil fuel markets is allocated to the Federal Tariff Service (FTS) and the Federal Anti-Monopoly Service (FAS).

Oil, natural gas and electricity markets are highly concentrated and remain dominated by government-owned companies, especially at the exploration and production stage. The oil sector is dominated by government-owned Rosneft (more than 40% of production and refining), with the remainder being split between public and private companies. The domestic and international distribution network is almost exclusively controlled by Rosneft, which also owns the retail network and retail companies for oil products. State-owned Gazprom controls all stages of natural gas exploration (72%), production (74.4%), processing (47.6%), distribution and exports, and is active in primary processing of oil and power generation. Slow liberalisation of the natural gas market and abolishment of the LNG export monopoly attracted some new entrants to the market (Henderson, 2012). Gazprom and subsidiaries are obliged to supply gas to all Russian customers in the regulated market segment (Gazprom, 2012), while non-Gazprom producers are not subject to price regulation.

The coal industry is privatised and relatively fragmented. The largest coal producer is SUEK (the Siberian Coal Energy Company), which in 2010 produced just under a third of total Russian coal production, of which about half is sold to electricity generators and a third is exported. Other major producers include the vertically-integrated steel companies such as Evraz, Mechel and Severstal.

Restructuring of the electricity market has brought new entrants and greater diversity of generation ownership. The transmission grid, and nuclear and hydro power generation remain under state control (IEA, 2013) Furthermore, government-owned enterprises continue to own or control over 60% of generation assets (*ibid.*).

Russia plans to expand the share of renewables to 2.5% of energy use by 2020, from the current 0.8%. Subsidies for the research and development of renewable energies will prioritise projects in line with a local content requirement, obliging solar, wind and hydro developers to use 20% of locally produced equipment (REW, 2014). Further, the country intends to reduce energy intensity of GDP by 40% by 2020 compared to 2008 through energy saving, improving energy efficiency and eliminating regulatory constraints, and via state subsidies and loan guarantees for efficiency improvement projects. Currently, progress towards reaching these targets is highly uneven (OECD, 2014).

Main energy policies influencing producer prices

Oil and oil product prices are not directly regulated, but are influenced via taxes and other means. As a response to growing domestic demand for refined oil products and fuel shortages, gasoline purchases were rationed in some regions in 2011 and 2012, combined with a sharp decrease in the gasoline export tax, introducing a de facto export ban (Henderson, 2012). In February 2011, the FAS fined some oil companies for over-charging, followed by an agreement with oil companies to temporarily freeze prices at the December 2011 level until May 2012 (Kojima, 2013). In addition, tax reforms of export and mineral extraction tax in October 2011 and January 2014 aimed at increasing refining complexity, to meet domestic motor and jet fuel demand.

Natural gas prices for the municipal, district heating and residential sectors, pipeline transmission and distribution charges are regulated by FTS at below-market levels, at less than one-fourth of the export price (Gazprom, 2014). While regulated domestic prices were meant to increase to export net-back[1] level by 2011, price increases for residential and municipal users have been capped to a yearly 15% until 2015 and remain at below market rates (Pirani, 2011). Regulated prices in regions remote from production areas are higher than prices in regions located closer to gas fields and large consumers. However, independent producers have been able to increase their market share in the unregulated market segment, industrial consumers and power generation (*The Moskow Times*, 2013; Kristalinskaya, 2014). Since 2006, Gazprom rations gas volumes available to industrial consumers at regulated prices in an undisclosed process, and is since 2007 allowed selling gas at commercial prices in the domestic market to new customers.

The mineral extraction tax is an important energy pricing instrument, and is imposed on the extraction of crude oil, natural gas and gas condensate. Tax rates vary by resource according to a world price factor and other coefficients, with the highest tax rate applying to the extraction of gas condensate. Non-Gazprom gas producers pay a lower tax rate on natural gas extraction. Significant tax expenditures arise from variation in MET tax rates, and concessions are given for a number of newly-developed on- and off-shore oilfields in specific regions.

The Russian wholesale electricity market is subdivided into an electric capacity market with different levels of price regulation and competition. The wholesale electricity market has been fully liberalised and prices have generally reflected movements in underlying supply-demand fundamentals and short-run marginal production costs, though prices are about one-third higher in the Europe and Urals prices zone, than in the Siberian one, due to congestion (IEA, 2013; Gore, 2011). The capacity market includes 27 zones, with prices regulated by FAS in 24 of them. Retail electricity prices are regulated by FTS at cost-plus prices for residential and small commercial customers, and cross-subsidised between customer classes, regions and generation technologies (IEA, 2013). Large-scale industrial and commercial electricity users can choose to buy electricity at unregulated prices from independent retailers, which now account for about one-fourth of the market (*ibid.*).

Export taxes apply to crude oil, oil products and natural gas with rates based on a formula according to price movements on the European oil market and individual field profitability. Tax exemptions and rebates apply among others to newly developed fields in East Siberia and the Caspian Sea, to encourage further development.

Structure of energy taxation

Companies in the downstream segment pay excise taxes on gasoline, naphtha, diesel and motor oils upon sales of excisable goods in the domestic market and imports. Tax rates vary depending on the category of excisable goods and are periodically adjusted by the tax authorities.

Exports are exempt from excise tax, so is ethanol for non-drinking purposes, as well as products which have been confiscated by customs and are subsequently resold. A producer of excisable goods can deduct excise tax paid on the purchase or import of the goods used in the production of these goods.

Russia applies VAT at 18%, with no exemption or reduced rates for energy products. Please refer to Section 1.4.3 for details on VAT policy. For the reasons discussed in Section 2.2, VAT is not included in the graphical profiles.

Energy use and taxation in Russia

Coupled with a sharp contraction in GDP growth linked to the collapse of the Soviet Union, total final energy consumption in Russia has decreased by about 50% between 1990 and 2000 and increased by 6% between 2000 and 2010.

As shown in the graphical profiles, transport makes up 14% of total energy use. In this category, 40% of energy used is gasoline, making up for 42% of emissions, 33% of energy and 28% of emissions is from natural gas and 20% of energy and 22% of carbon emissions is from diesel. Remaining transport fuels are fuel oil, LPG and aviation fuels. Of these fuels, gasoline, fuel oil and diesel are subject to the excise tax, with the highest rate applying to gasoline, followed by fuel oil and diesel. Naphtha is taxed as well, but is used as an intermediate product in the refining process.

Energy for heating and process use accounts for 52% of total energy use, with natural gas accounting for 62% of energy use (48% of carbon emissions) in that category. Coal and coal gases account for 22% of energy use and 37% of emissions, while diesel, LPG, crude oil and other oil products make up for 14% of energy used and 13% of carbon emissions in this category. Within heating and process use category, only diesel is taxed, under the excise tax.

The main constituents of electricity are natural gas (51% of inputs to electricity generation and 58% of emissions from electricity inputs), nuclear (21% of inputs to electricity generation) and coal or coal gases (18% of inputs and 38% of carbon emissions). Remaining electricity is generated from hydro as well as other renewables, combustibles and waste. Neither electricity fuels or consumption are taxed.

Reported tax expenditures and rebates relevant for graphical profiles

None.

Information on tax rates and bases, tax expenditures energy use, and conversion rates were obtained from the sources detailed in Annex A, from Government websites, and from consultation with national officials.

In addition, country-specific sources were used as listed in the references for this chapter.

Figure 6.1. **Taxation of energy in Russia on an energy content basis**

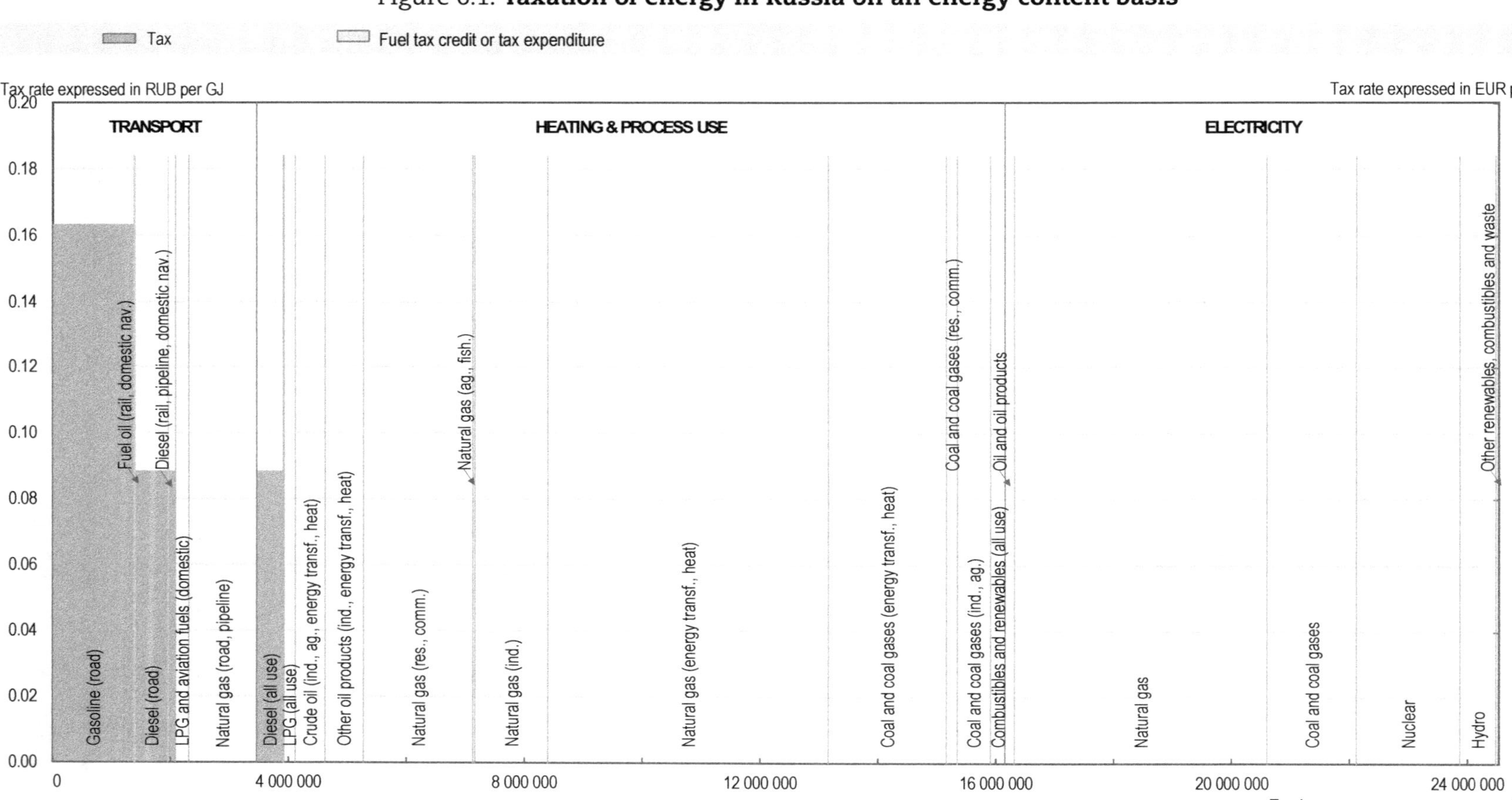

Source: OECD calculations. Tax rates are as of 1 April 2012; energy use data is for 2009 from IEA (2014), *IEA World Energy Statistics and Balances* (database), Doi: http://dx.doi.org/10.1787/data-00513-en.

StatLink http://dx.doi.org/10.1787/888933205909

Figure 6.2. **Taxation of energy in Russia on a carbon content basis**

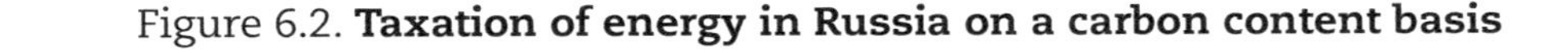

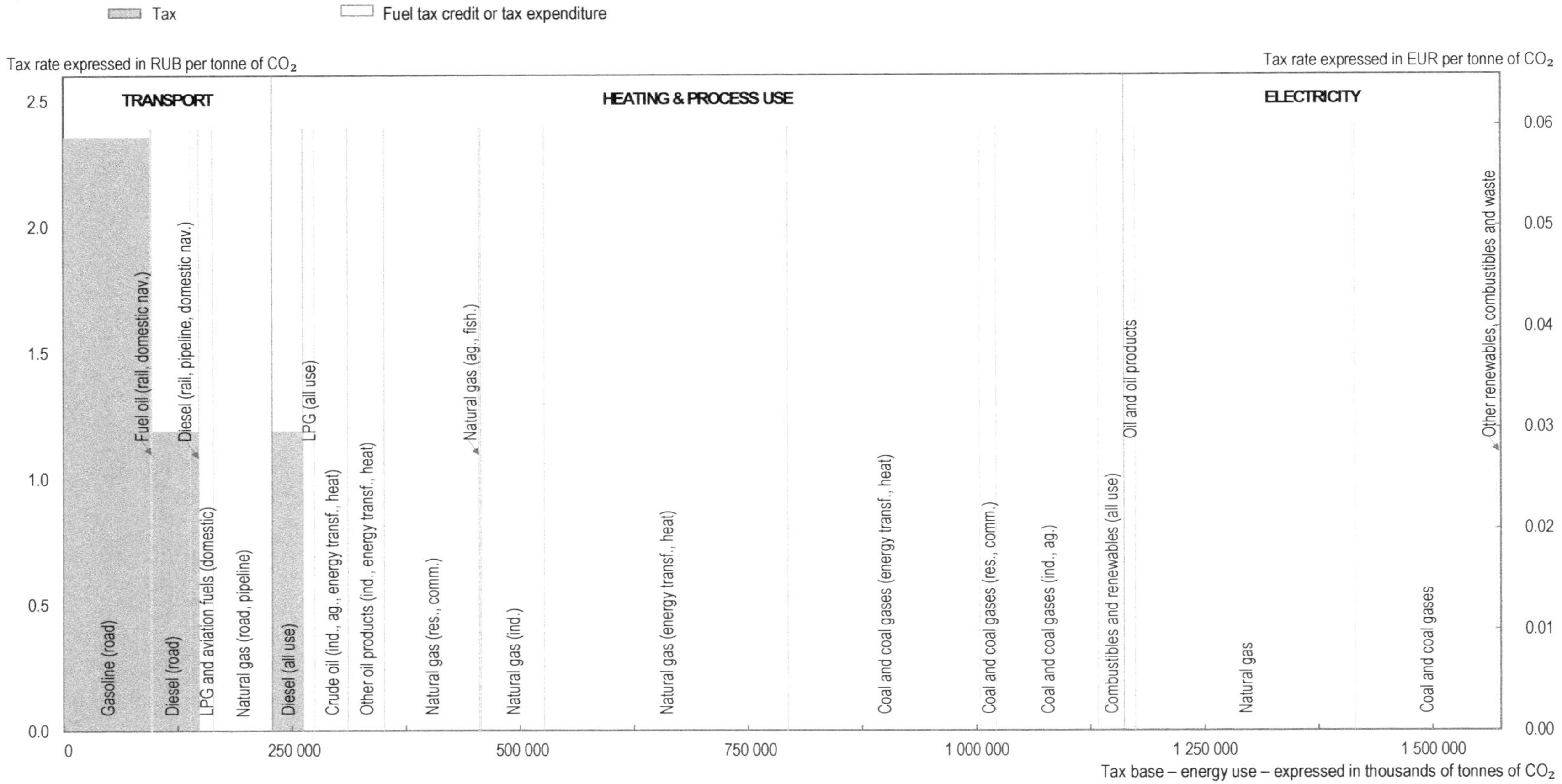

Source: OECD calculations. Tax rates are as of 1 April 2012; energy use data is for 2009 from IEA (2014), *IEA World Energy Statistics and Balances* (database), Doi: http://dx.doi.org/10.1787/data-00513-en.

StatLink http://dx.doi.org/10.1787/888933205916

Note

1. The underlying calculation for net-back prices takes Gazprom's export price in Europe less transport costs from the Russian border, less export tax (30% for natural gas), less the transport from the market in Russia (generally taken to be Moscow).

References

Banham, R. (2012), "Opportunity Knocks", World Coal, March 2012, *www.deloitte.com/assets/Dcom-Russia/Local%20Assets/Documents/Energy%20and%20Resources/dttl_Opportunity-knocks_Russell-Banhans-article_March.pdf*.

Cooke, D. (2013), "Russian Electricity Reform 2013 Update: Laying an Efficient and Competitive Foundation for Innovation and Modernisation", *International Energy Agency Insights 2013, OECD/OEA Publishing*, Paris.

Deloitte (2012), "Tax and Legal Guide to the Russian Oil and Gas Sector", 2012 ZAO Deloitte and Touche CIS.

Ernst and Young (2013a), "Global Oil and Gas Tax Guide", *2013 EYGM Limited*.

Ernst and Young (2013b), "A Switch from the 60-66 System to 55-61 in the Russian Oil Industry", *EY Oil and Gas Alert, October 2013*, 2013 EYGM Limited.

Ernst and Young (2011), "Russian Federation Oil Tax Reform", *EY Oil and Gas Tax Alert*, September, 2011 EYGM Limited.

Fattouh, B. and J. Henderson (2012), "The Impact of Russia's Refinery Upgrade Plans on Global Fuel Oil Markets", WPM 48 *The Oxford Institute for Energy Studies.*

Gazprom (2014), "Competitive Environment – Prerequisite for Further Russian Gas Market Development", *Gazprom News*, March, *www.gazprom.com/press/news/2014/march/article187091*.

Gerasimchuk, I. (2012), "Fossil Fuels – At What Cost? Government Support for Upstream Oil and Gas Activities in Russia", *Report for WWF Russia and the Global Subsidies Initiative (GSI) of the International Institute for Sustainable Development (IISD)*, Moscow-Geneva.

Gore, O., S. Viljainen, M. Makkonen and D. Kuleshov (2012), "Russian Electricity Market Reform: De-Regulation or Re-Regulation?", *Energy Policy* 41, (2012) 676-685.

Henderson, J. (2012), "Evolution in the Russian Gas Market – The Competition for Customers", *The Oxford Institute for Energy Studies*, January 2013.

Henderson, J. (2011), "Domestic Gas Prices in Russia – Towards Export Netback?", *The Oxford Institute for Energy Studies*, November.

IEA (2014), "Extended World Energy Balances", *IEA World Energy Statistics and Balances* (database), Doi: *http://dx.doi.org/10.1787/data-00513-en*.

Kuleshov, D., S. Viljainen, S. Annala and O. Gore (2012), "Russian Electricity Sector Reform: Challenges to Retail Competition", *Utilities Policy* 23, (2012) 40-49.

Kristalinskaya, S. (2014), "Gas Tariffs: Independents on the Offensive", *Oil and Gas Journal Eurasia*, *www.oilandgaseurasia.com/en/news/gas-tariffs-independents-offensive*.

Ministry of Energy (n.d.), Ministry of Energy of the Russian Federation, Description, *http://government.ru/en/department/85/about*.

Ministry for Economic Development of the Russian Federation (n.d.), "Department for State Governance of Tariffs, Infrastructural Reforms and Efficient Energy Use", *www.economy.gov.ru/wps/wcm/connect/economylib4/en/home/about/stucture/depGostarif*.

OECD (2014), *Economic Survey Russian Federation*, OECD Publishing, Paris, Doi: *http://dx.doi.org/10.1787/eco_surveys-rus-2013-en*.

Oil and Gas Eurasia (2014), "Gazprom Board Reluctant to Continue Subsidizing Russian Economy", 23 April, *www.oilandgaseurasia.com/en/news/gazprom-board-reluctant-continue-subsidizing-russian-economy*.

Pirani, S. (2013), "Consumers as Players in the Russian Gas Sector", Oxford Energy Comment, *The Oxford Institute for Energy Studies.*

Renewable Energy World, REW (2014), "Putin Strengthening China-Russia Ties for Renewable Energy Development", *www.renewableenergyworld.com/rea/news/article/2014/10/putin-strengthens-russias-chinese-renewable-energy-ties*.

South Africa

This chapter discusses the energy landscape, pricing policies, and the structure of energy taxation in South Africa and presents graphical profiles of energy use and taxation in South Africa in both energy and carbon terms (Figures 7.1 and 7.2, respectively).

South Africa has large coal reserves and produces almost all coal used domestically. It exports around 25% of production. South Africa's energy consumption is dominated by coal, particularly bituminous coal, at 71% of total primary energy supply. It is particularly dominant in electricity production. The coal sector is largely privately owned, and both international and domestic companies operate in this sector.

South Africa refines oil products domestically, covering 81% of domestic consumption in 2009, with the remainder being imported. South Africa produces a small amount of natural gas (around 20% of natural gas used in 2009) and imports the remainder, primarily from Mozambique. Almost all natural gas is used in energy transformation processes. Non-fossil fuel sources of energy include renewables and biomass. Renewables are used in electricity generation to a small extent. Biomass is used by residential households in small quantities.

The Central Energy Fund is responsible for the financing and promotion of coal acquisition and the manufacture of synthetic fuels from coal and the acquisition, generation, manufacture, marketing or distribution of other forms of energy (CEF, 2014). It is a State-owned company under the responsibility of the Minister of Energy and is the parent organisation for several other State-owned corporations, notably PetroSA and the Petroleum Agency of South Africa (PASA).

PetroSA, a State-owned company, is responsible for almost all oil produced in South Africa (EIA, 2014), but only covered 6% of refining, distribution and retail in 2012 (Competition Tribunal of South Africa, 2012). Sasol, a private, publicly traded company, is active in oil refining, and operates a major coal-based synthetic fuel plant which generated around 40% of crude oil in 2009 (Sasol, 2014). The distribution and retail of oil products is more diversified and several international and domestic companies are involved, with BP, Shell, Chevron, Total and Engen as the main players (EIA, 2014). The oil sector is regulated by PASA. PASA also regulates the exploration and production of natural gas, and PetroSA is responsible for all domestic production. Natural gas from Mozambique is imported through the Sasol Petroleum International Gas pipeline, which is owned by a joint venture between Sasol, the South African government and the Mozambican government.

The National Energy Regulator of South Africa (NERSA) regulates the electricity sector, as well as oil and natural gas pipeline industries. It is also responsible for electricity pricing. Around 95% of electricity is generated by the State-owned electricity company Eskom. In the White Paper on Renewable Energy (2003), South Africa has set itself a target to supply about 5% of electricity generation from renewable sources by 2013, mainly from biomass, wind, solar and small-scale hydro. In 2009, that share was at below 1%, according to IEA (2011).

Key energy policies affecting producer prices

South Africa regulates the prices of some refined oil products through two main mechanisms – a basic fuel price (import parity pricing formula) for gasoline, diesel and kerosene, and maximum prices for certain fuels and piped natural gas. Electricity prices are also regulated. The prices of coal and natural gas are not regulated, although domestic coal prices remain below international prices.

The prices of gasoline, diesel and kerosene are set by reference to a basic fuel price. The basic fuel price is determined each month by the Department of Energy, based on international market prices for each fuel. Domestic wholesale and transport costs are added to the basic fuel price, together with a number of taxes and levies (discussed below) to arrive at the end-user price. The cumulative difference between the Basic Fuel price determined monthly and the daily Basic Fuel Price is recorded in the Slate Account. When the cumulative negative balance of the slate account exceeds ZAR 250 million, a Slate Levy applies to these fuels to finance the difference. The amount of the levy depends on the balance of the Slate Account.

The government also sets maximum retail prices for LPG, kerosene (illuminating paraffin) and piped natural gas. For LPG, a maximum ex-refinery price is set, together with maximum retail prices by region (with the maximum retail margin set at 15% of wholesale price). The maximum price for kerosene is set by reference to the Basic Fuels Price, a wholesale margin, service and router differentials and transport costs.

Regulation of the price of coal ended in 1987. However, due to long-term contracts between Eskom and coal manufacturers, the price of coal used in electricity generation is still well below market and export coal prices.

The maximum price for piped gas is based on a weighted basket of prices for alternative fuels, including coal, diesel, electricity, fuel oil and LPG.

In the electricity sector, Eskom supplies approximately 95% of all electricity. The tariffs Eskom can charge for electricity are set by the National Energy Regulator of South Africa, and are differentiated across different use groups (for example, residential, commercial, industrial and public lighting) and in urban or rural areas. In the residential sector, higher-use consumers pay higher prices, subsidising lower-use groups. Low-income households are provided with a free basic electricity entitlement of at least 50 kWh per month. The government reports an expenditure of ZAR 675 per low-income household (OECD, 2014).

Structure of energy taxation

South Africa applies fuel taxes to gasoline, diesel, kerosene, LPG and fuels for domestic aviation. Other fuels, including other oil products, coal, and natural gas are untaxed when used for transport or heating and process purposes. The Electricity Levy, a consumption tax on electricity, applies to electricity generated from non-renewable resources and to nuclear-generated electricity.

Fuel taxes that apply to gasoline, diesel, kerosene and LPG include:

- **The General Fuel Levy:** this levy was introduced in 1983 and applies to gasoline and diesel. Revenue from this levy is used to fund the government's general expenditure programmes, including the construction and maintenance of roads and support of public transport, and around one-third is shared with metropolitan municipalities. Rates are adjusted annually. Refunds from this levy apply to diesel used for certain purposes in primary production sectors. Full rebates apply to offshore diesel (e.g. offshore mining, research, patrols, or fisheries) and to diesel used in electricity generation, where the plant has an installed capacity of over 200 MW. A partial rebate applies to onland diesel use (including agriculture, forestry and mining), at a rate of 40% of the General Fuel Levy on 80% of eligible diesel purchases. The refunds are to promote competitiveness in primary production sectors and to provide relief for off-road use of gasoline and diesel.

- **Road Accident Fund Levy:** Introduced in 1997, this levy is used to provide cover for injuries or death from road accidents. It applies to gasoline and diesel, and rates are adjusted annually. Refunds from this levy apply to diesel used for certain purposes. A full rebate applies to diesel used offshore and in electricity generation (under the same conditions as described in relation to the General Fuel levy); and to diesel used in rail. Onland diesel use is subject to a full refund of the Road Accident Fund levy for 80% of the diesel purchases subject to the levy. These refunds also exist to ensure competitiveness of the rail industry and for fairness purposes, to exempt non-road users from a levy designed to deal with the consequences of motor vehicle accidents.
- **Customs and Excise Levy:** This levy applies to petrol, diesel and biodiesel. It was introduced in 1983 and the rate has remained unchanged since. No exemptions or refunds apply to the Customs and Excise Levy.
- **Illuminating Paraffin Diesel Marker Levy:** The illuminating paraffin diesel marker applies to diesel and to kerosene. It was introduced to finance expenses involved in marking illuminating paraffin (kerosene) to prevent it from being illegally mixed with diesel. It was introduced in 1999 and the rate was reduced in 2004 from ZAR 0.022 to 0.0001 per litre. No exemptions or refunds apply to the illuminating paraffin diesel marker levy.
- **Petroleum Pipelines Levy:** This levy was introduced in 2007 to meet the general administrative costs of the Petroleum Pipelines Regulatory Authority. Until 2013, it applied at a rate of ZAR 0.0015 per litre, which was then increased to ZAR 0.0029 per litre. No exemptions or refunds apply to the Petroleum Pipelines Levy.
- **Slate Levy:** The Slate Levy has applied to gasoline and diesel since 2009. The levy finances a negative balance in the Slate account beyond ZAR 250 million, which records the difference between daily Basic Fuel Prices and the monthly Basic Fuel Price which is used in the pricing structure for diesel and gasoline. The amount of the levy is determined by the self-adjusting Slate Levy Mechanism based on the cumulative balance of the Slate account.
- **Incremental Inland Transport Recovery Levy**: Introduced in 2008, this levy applies to gasoline and diesel at a rate of ZAR 0.03 per litre. It was implemented to finance alternative modes of transports for regulated petroleum products and jet fuel from the coast to a set of nominated pipeline zones, due to capacity constraints on the pipeline which transports these fuels from the coast to the inland. No exemptions or rebates exist for this levy. This levy was abolished in April 2014.
- **Demand Side Management Levy:** This levy applies to 95 octane gasoline used in inland magisterial districts at a rate of ZAR 0.1 per litre. It was implemented to deter unnecessary use of high octane gasoline in these areas. In 2012, 47% of gasoline use was of 95 octane gasoline, of which 27% was subject to the Demand Side Management Levy (12.8% of all gasoline). The proportion of gasoline subject to the levy is shown separately in the graphical profiles.

The rates of these levies shown in the graphical profiles, and the fuels and uses they apply to, are summarised in Table 7.1.

Table 7.1. **Fuel taxes and levies on gasoline, diesel, LPG and kerosene in South Africa**

Rate as at 4 April 2012 (ZAR cents per litre)	Gasoline		Diesel				LPG and kerosene
	General rate	95 octane, inland use	General rate	Rail	Onland[1] (mining, ag., forestry)	Offshore (fishing, coasting, mining, patrol)	
General Fuel Levy (GFL)	197.50	197.50	182.50	182.50	118.5	–	–
Road Accident Fund Levy	88.00	88.00	88.00	–	–	–	–
Customs and Excise SACU pool	4.00	4.00	4.00	4.00	4.00	4.00	–
Illuminating Paraffin Diesel Marker Levy	–	–	0.01	0.01	0.01	0.01	0.01
Petroleum Pipelines Levy	0.15	0.15	0.15	0.15	0.15	0.15	–
Slate Levy	15.36	15.36	15.36	15.36	15.36	15.36	15.36
Incremental Inland Transport Recovery Levy	3.00	3.00	3.00	3.00	3.00	3.00	–
Demand-side Management Levy	–	10.00	–	–	–	–	–
Total	308.01	318.01	293.02	205.02	141.02	22.52	15.37

1. The taxes applying to onland and offshore diesel only apply to 80% of eligible purchases, not to the total diesel used in these sectors.

Source: OECD calculations based on data provided by national officials.

In addition to these taxes, South Africa levies an aviation fuel levy of ZAR 0.13 per litre on aviation fuel for domestic use, which is used to fund the Civil Aviation Authority. It is not payable on exported aviation fuel, international aviation, or flights for which the Passenger Safety Charge is payable. This levy is also shown in the graphical profiles.

The Environmental Levy on Electricity applies to the consumption of electricity generated from non-renewable sources and nuclear plants. The levy, which applies at a rate of ZAR 0.035 per kWh, does not apply to electricity generated from hydro, renewables, or biomass. Power plants that have an installed capacity of less than 5 MW are exempt from the levy. This levy is also shown in the graphical profiles.

In 2010, the Government announced a carbon tax, which would apply to fuels based on their carbon content at a rate of ZAR 120 per tonne of CO_2. It will cover emissions from fuel combustion and non-energy industrial use of fuel, amounting to approximately 80% of total emissions. The introduction of the tax has been delayed until 2016 to allow alignment with other environmental policies under development.

The standard VAT rate of 14% applies to most energy products and to the consumption of electricity, although gasoline, diesel and illuminating paraffin (kerosene) are zero-rated for VAT purposes. See Section 1.4.3 for a fuller description. Neither the standard or differential VAT rates are shown in the graphical profiles for the reasons discussed in Section 2.2.

Energy use and taxation in South Africa

As can be seen from the graphical profiles, energy used in transport accounted for 13% of total energy use and 10% of carbon emissions from energy use. Within the transport category, gasoline is the highest taxed fuel and accounts for 54% of energy use and 46% of carbon emissions from transport energy. Diesel for road use is taxed at a lower rate and is the second largest energy source in this category, representing 39% of energy use and 41% of emissions. Diesel for rail use accounts for less than 1% of transport energy and emissions. The remainder of energy used in the transport category is used in aviation (6%) which are taxed under the aviation fuel levy, although at lower rates than gasoline and diesel for road transport.

44% of total energy use is used for heating and process purposes, generating 47% of total emissions. Oil products account for 8% of energy use in this category (6% of emissions), of which diesel, gasoline and kerosene used in this category are taxed. Partial refunds apply to onland use of diesel. Heating and process energy is dominated by coal (66% of energy used for heating and process purposes and 65% of carbon emissions from heating and process energy use), which is untaxed. The remainder of heating and process energy is derived from biomass and charcoal (24% of energy and 28% of carbon emissions from energy use), natural gas (2% and 1% of emissions) and solar energy (0.1% of energy use).

The remaining 43% of energy use is used to generate electricity, generating 43% of total CO_2 emissions from energy use. Bituminous coal represents 94% of energy used in electricity generation and is taxed under the Environmental Levy on Electricity. Nuclear energy (5% of energy use) is also taxed under this levy. A small amount of diesel (0.02% of energy used in electricity generation in 2009) is also used, with the remainder being generated from biomass, hydro and solar energy (0.4% in total).

Reported tax expenditures and rebates

The refund of the General Fuel Levy and the Road Accident Fund levy are shown in the graphical profiles under the subcategories "Diesel (rail)" and "Diesel (ag., mining., fish.)". Refunds applying to electricity generation are too small to be shown but would appear under the subcategory "Diesel" in the Electricity category.

These refunds are shown in the graphical profiles as refunds against the full rate of both levies, rather than as tax expenditures. This is because both the General Fuel Levy and the Road Accident Fund Levy are designed to apply to road use of fuels and to address externalities associated with road fuel use. On this basis, refunds of these charges for non-road users are not considered tax expenditures.

Key assumptions and caveats

Power plants that have an installed capacity of less than 5 MW are exempt from the Environmental Levy on Electricity. As 95% of all electricity is produced by Eskom, we have assumed that no power plants fall under this exemption.

Electricity plants that have an installed capacity of greater than 200 MW are able to claim back the General Fuel Levy and the Road Accident Levy on diesel used in electricity production. Only the Ankeling and Gourikwa plants, operated by Eskom, are able to claim this refund. In 2009, the amount of diesel used to produce electricity accounted for 0.02% of electricity generation fuels, so the refunds are not distinguishable in the graphical profiles.

Information on tax rates and bases, tax expenditures on energy use, and conversion rates were obtained from the sources detailed in Annex A, from Government websites, and from consultation with national officials.

In addition, country-specific sources were used as listed in the references for this chapter.

Figure 7.1. **Taxation of energy in South Africa on an energy content basis**

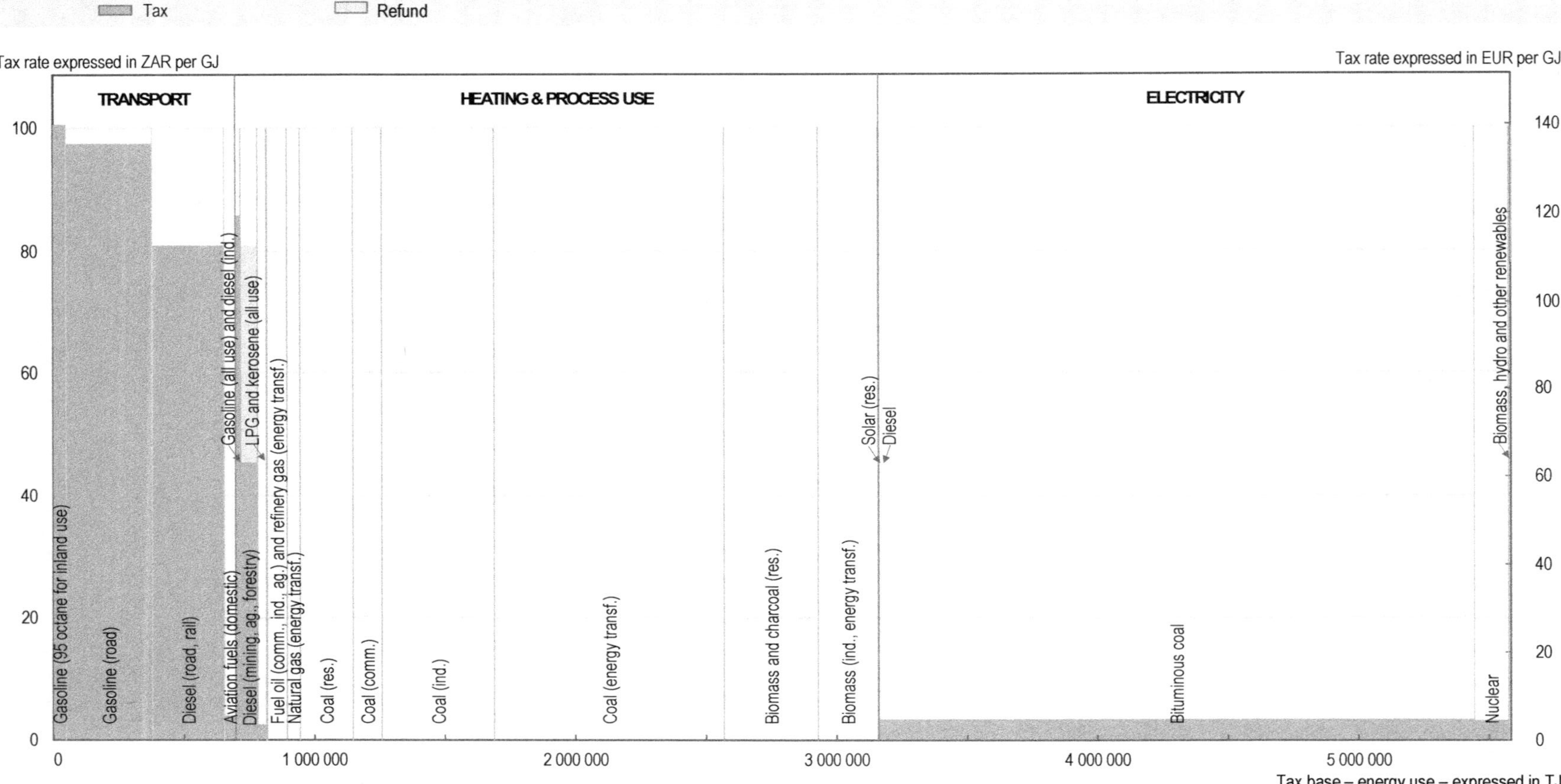

Source: OECD calculations. Tax rates are as of 1 April 2012; energy use data is for 2009 from IEA (2014), *IEA World Energy Statistics and Balances* (database), Doi: http://dx.doi.org/10.1787/data-00513-en.

StatLink http://dx.doi.org/10.1787/888933205928

Figure 7.2. **Taxation of energy in South Africa on a carbon content basis**

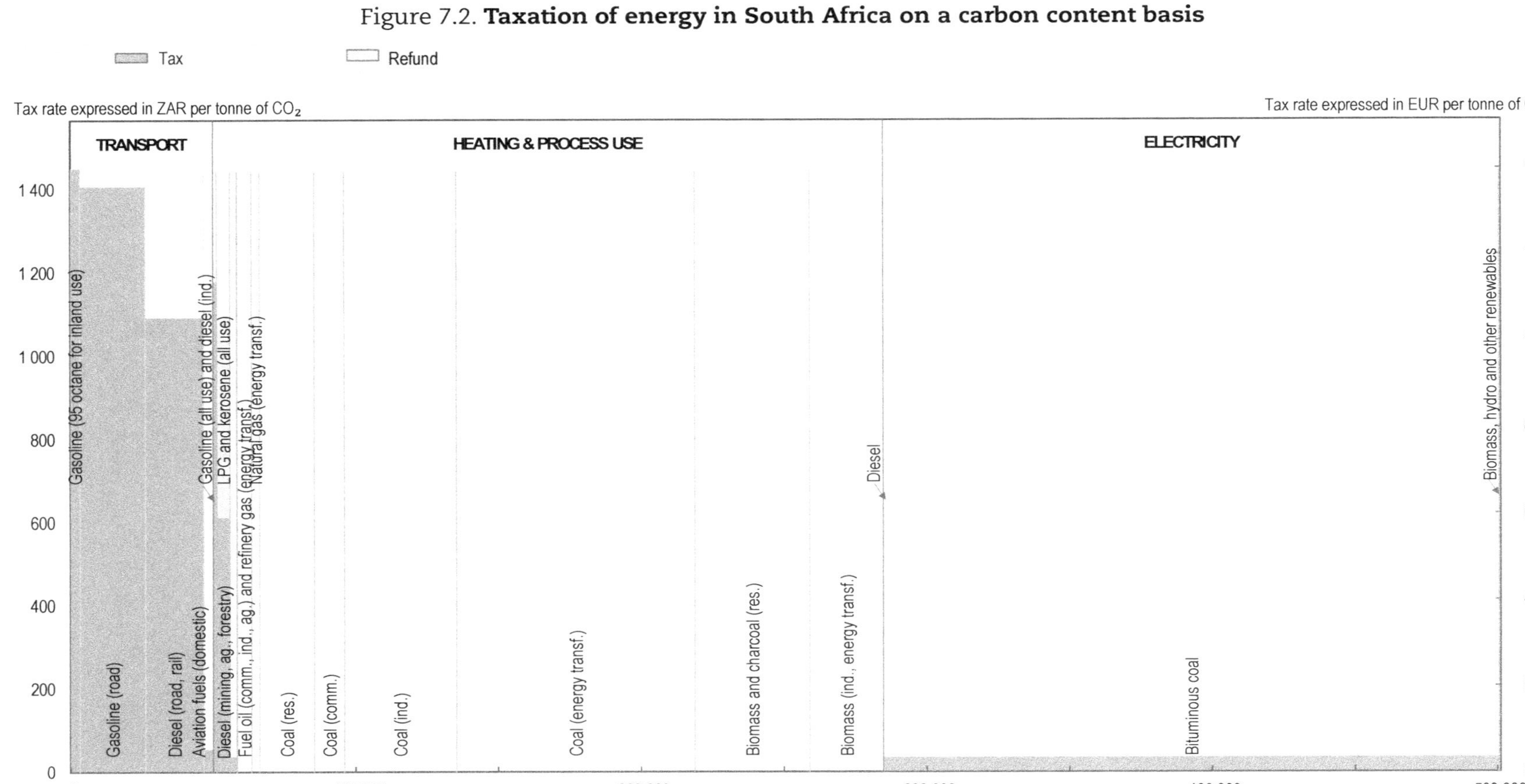

Source: OECD calculations. Tax rates are as of 1 April 2012; energy use data is for 2009 from IEA (2014), *IEA World Energy Statistics and Balances* (database), Doi: http://dx.doi.org/10.1787/data-00513-en.

StatLink http://dx.doi.org/10.1787/888933205933

References

Competition Tribunal of South Africa (2012), "Petroleum Oil and Gas Corporation of South Africa", *www.saflii.org/za/cases/ZACT/2012/61.pdf.*

Department of Energy (2014a), "Working Rules to Set the Monthly Maximum Retail Price for Liquefied Petroleum Gas (LPG)", *www.energy.gov.za/files/policies/WORKING_RULES_2010.pdf.*

Department of Energy, (2014b), "Petroleum Price Structure", *www.energy.gov.za/files/esources/petroleum/petroleum_pricestructure.html.*

Department of Energy (2008), "Rules to Administer the Self-Adjusting Slate Levy Mechanism", *www.energy.gov.za/files/esources/pdfs/energy/liquidfuels/slate%20levy.pdf.*

Eskom (2012), "Tariffs and Charges Booklet 2012-13", *www.eskom.co.za/CustomerCare/TariffsAndCharges/Documents/ESKOM%20TC%20BOOKLET%202012-13%20FINAL%203.pdf.*

IEA (2014), "Extended World Energy Balances", *IEA World Energy Statistics and Balances* (database), Doi: *http://dx.doi.org/10.1787/data-00513-en.*

Kohler, M. (2013), *"Differential Electricity Pricing and Energy Efficiency in South Africa"*, *www.econrsa.org/system/files/publications/working_papers/working_paper_396.pdf.*

NERSA (2011a), "Methodology to Approve Maximum Prices of Piped-Gas in South Africa", *www.nersa.org.za/Admin/Document/Editor/file/Piped%20Gas/Pricing%20and%20Tariffs/Pricing%20Decisions/Current/Methodology%20%20to%20approve%20Maximum%20Prices%20for%20Piped-Gas%20in%20South%20Africa.pdf.*

NERSA (2011b), "Methodology to Approve Maximum Prices of Piped-Gas in South Africa", *http://www.nersa.org.za/Admin/Document/Editor/file/Piped%20Gas/Pricing%20and%20Tariffs/Pricing%20Decisions/Current/Methodology%20%20to%20approve%20Maximum%20Prices%20for%20Piped-Gas%20in%20South%20Africa.pdf.*

NERSA (2006), "2006 Electricity Supply Statistics for South Africa", *www.nersa.org.za/Admin/Document/Editor/file/News%20and%20Publications/Publications/Current%20Issues/Electricity%20Supply%20Statistics/Electricity%20supply%20statistics%202006.pdf.*

Sasol (2014), "Sasol Facts 12/13", *www.sasol.com/extras/sasol-facts-pres-2.*

Statistics South Africa (2012), "Electricity Generated and Available for Distribution (Preliminary) December 2012", *http://beta2.statssa.gov.za/publications/P4141/P4141December2012.pdf.*

South African Department of Energy (2003), "Renewable Energy Overview", *www.energy.gov.za/files/renewables_frame.html.*

South African Revenue Service (2014a), "Excise External Directive: Refunds", *www.sars.gov.za/AllDocs/OpsDocs/Policies/SE-FS-19%20-%20Refunds%20-%20External%20%20Directive.pdf.*

South African Revenue Service (2014b), "Environmental Levy on Electricity Generated in South Africa", *www.sars.gov.za/AllDocs/OpsDocs/Policies/SE-EL-04%20-%20Environmental%20Levy%20on%20Electricity%20Generated%20in%20South%20Africa%20-%20External%20Standard.pdf.*

South African Revenue Service (2013), "Guide to Excise Duties and Levies", *www.sars.gov.za/AllDocs/OpsDocs/Guides/SE-FS-09%20-%20Excise%20Duties%20and%20Levies%20-%20External%20Guide.pdf.*

South African Government (2015), "Customs and Excise Tariff Act", Schedule 6, *www.sars.gov.za/AllDocs/LegalDoclib/SCEA1964/LAPD-LPrim-Tariff-2012-19%20-%20Schedule%20No%206.pdf.*

ANNEX A

Methodology and data sources

The graphical profiles show the composition of energy use in each country covered and the effective rate of tax on various segments of energy use. Both energy use and tax rates are shown alternately in terms of energy content and carbon content.

This annex provides a detailed overview of the methodology, assumptions, and data sources underlying the graphical profiles. Details specific to a particular country are found in the relevant country chapter.

Tax base-energy use

The tax base presented on the horizontal axis of the maps includes all energy use within a country including: the use of energy by end-use residential and business consumers; the use of energy in the generation of electricity (shown in the electricity category) and of heat for sale; and the net energy that is used to transform energy into other forms, or to distribute energy. Given the focus on consumption within each country, imports of energy are included (with the exception of fuels used to produce electricity, detailed below) whereas exports of energy to other countries are excluded. Fuel used in international aviation and international maritime shipping is similarly excluded.

Energy consumption is divided into three main categories: transport fuel use, heating and process fuel use and electricity. Each of these categories has then been disaggregated into a number of subcategories reflecting how tax bases are defined in the particular national tax system. The subcategories therefore differ from country to country and may reflect, variously, the nature of the fuel, the use of the fuel or the user of the fuel.

Within the main categories, transport is relatively straightforward, with a few major fuels (motor gasoline and diesel) and some smaller items, like biofuels. Heating and process fuel use (comprised from IEA data on industrial, commercial and residential use) is more complex because the same fuels can be taxed very differently depending on whether they are used for industrial processes or for heating and, in some countries, depending on household location or size. The heating and process fuel use category also includes the net energy used to transform energy from one form to another – e.g., in oil refineries, which transform crude oil into products like motor gasoline and diesel, or in merchant heating plants that burn fuel to produce heat for sale. Net energy use has been identified by calculating the difference (in terms of energy content or carbon emissions) between the energy that has been used by each type of energy transformation plant and the energy output by that plant as another fuel available for use by an end-use consumer. In the IEA data, fuels used by industry for transport off business premises (for example, for transit of goods on public roads) are included in the transport sector, whereas fuels used as propellant on business premises are listed in the heating and process category.

Treatment of electricity

Electricity is a form of secondary energy that is generated using some other source of primary energy (e.g., coal, natural gas or hydro). While relatively few countries tax fuel that is used to generate electricity, many countries have a tax on electricity consumption that may be analysed as an indirect tax on the fuels used to generate the electricity. Given the focus on fuel combustion, therefore, the paper includes electricity consumption but "looks through" to the underlying primary energy used to generate it (provided both the electricity use and the generation take place within the country). All fuels used to produce electricity for domestic consumption, including energy industry own use, are included.

The graphical profiles show the domestic consumption of electricity that has been produced domestically. They exclude both electricity generated abroad that is imported and electricity generated domestically which is exported. This is for two reasons:

- In the case of electricity imports, the IEA data does not indicate the producing country or the underlying primary energy source used to generate the electricity.
- Since the maps are based on consumption, it would not seem appropriate to attribute exported electricity to the exporting country.

Given that electricity is generally fungible on the grid, and the generation source of exported electricity cannot be readily identified, the fuels used to generate electricity for export have been assumed to be in the same proportion as for total domestic generation.

"Looking through" domestic electricity consumption to the domestic energy sources used to generate that electricity involves several adjustments. Since the graphs illustrate taxes on the consumption of electricity as an indirect tax on the energy used to produce that electricity, it is necessary to estimate the amount of fuel used to generate the electricity consumed by each user and the related CO_2 emissions.

The energy value of the primary fuel used to generate a unit of electricity for consumption by end users is larger than the energy value of that unit of electricity. This is because there are inefficiencies in converting primary fuels into electricity as well as substantial transmission losses in sending electricity from the place of generation to the place of consumption. Therefore, the energy value of electricity consumed is grossed up to show the energy value of all the primary energy used to generate the electricity.[1]

Given the fungibility of electricity, it is assumed that electricity consumed by each user reflects on a *pro rata* basis the national distribution of generation sources. This amounts to an assumption that all end users use the same mix of electricity. In a similar way to energy use, tax rates on electricity consumption expressed in terms of energy value are adjusted, but in a downward direction, taking into account that the effective tax rate on the fuels used to produce electricity is lower than the nominal tax rate on the energy value of electricity consumed because of energy losses in generation and transmission.

After the adjustments, as with other types of energy, the electricity category in the graphical profiles is divided into sub-categories that reflect the way in which electricity is taxed in the particular country. Electricity is therefore shown in the profiles in one of two ways, by user or by fuel:

Table A.1. **Approaches to electricity in maps of energy use**

	Electricity shown by user of electricity	Electricity shown by fuels used to generate electricity
Tax base	Energy or CO_2 value of the average mix of fuels used to generate the electricity consumed by each user.	Energy or CO_2 value of each type of fuel used to generate electricity.
Tax rate	Effective tax rate on fuels used to generate the electricity consumed by those users.	Effective tax rate on each fuel used to generate electricity.

Data sources

Information on energy consumed in each country – the potential tax base – was obtained from the *Extended World Energy Balances* ("EWEB") (IEA, 2011a). *Taxing Energy Use – A Graphical Analysis* (OECD, 2013) shows data on energy use for 2009, which was the latest year for which data in spreadsheet form was available when the analysis was undertaken. To ensure consistency with this earlier analysis, analysis in this publication is based on the same data as in *Taxing Energy Use – A Graphical Analysis* (OECD, 2013). Data for 2009 is used essentially as a proxy for current consumption.

Supplementary information on energy use data for individual countries has been used where it is necessary to divide one of the categories in the EWEB to more accurately reflect a country's tax settings. This was necessary, for example, in cases where a country applies a lower tax rate in more remote or less developed areas of the country.

Tax rates

On the vertical axis, the graphical profiles show the tax rates and related tax expenditures that apply to energy use as at 1 April 2012 (except for AUS and BRA, which are shown as at 1 July 2012 and ZAF, which is used at 4 April 2012). The taxes covered are those that are levied selectively on some or all energy products, either as a per-unit tax on a physical basis such as volume or weight (which in some cases may be set to reflect energy or carbon content), or as a percentage of the sales price (*ad valorem* taxes).

In contrast to *Taxing Energy Use – A Graphical Analysis* (OECD, 2013), *ad valorem* taxes (i.e. not including VAT) and related tax expenditures set by reference to the value of products are included into the graphical profiles. This is since in several of the selected partner economies, non-VAT taxes on energy products levied as a percentage of the sales price are more commonly used (for example, in Argentina, India and Indonesia). Where this was the case, price information has been used to translate *ad valorem* taxes into per-unit tax rates. Where possible, average fuel prices of the six months before and after 1 April 2012 – the date for which tax rates are generally shown – are used (October 2011 until September 2012). Where this information was not available, shorter samples of fuel prices are used. Converting *ad valorem* taxes into per-unit rates allows the calculation of effective tax rates on energy and carbon terms across different bases (for example, effective tax rates on transport fuels), but the calculated unit taxes are contingent upon observed prices.

Taxes that apply to a very broad range of goods (such as value-added and retail sales taxes) are not included in the graphical profiles. On the other hand, where an energy product is subject, for example, to a concessionary rate of VAT, the concession would affect relative energy prices. In order to gauge to what extent such VAT rate differentiation takes place, VAT rates and concessionary VAT rates on energy products are discussed in Section 1.4.3. Also excluded are taxes that that may be related to energy use but that are not imposed directly on the fuel (such as vehicle taxes, road user or billing charges). Tax expenditures on bases other than energy consumption (such as renewable energy production or investment in equipment used for that purpose) are similarly excluded.

Ideally, the graphical profiles would show the impact of all policy measures influencing producer prices of energy. It was not possible, however, for the purposes of this study to obtain information on all relevant prices for all products and users charged in the countries considered, as well as on those prices which would apply in the absence of these policies. Consequently, *Taxing Energy Use – A Graphical Analysis* (OECD, 2013), as well as this report, consider the impact of energy taxes on the user price of energy from a common base of zero, reflecting an implicit view that taxes are added to producer prices of energy which align broadly with marginal production costs. Where price support measures, production taxes or export taxes exist and these measures differ among countries, this common base no longer applies. Some of the countries analysed in both this report and in *Taxing Energy Use – A Graphical Analysis* (OECD, 2013) apply price support measures, such as price controls, regulations or certain forms of support to fuel producers or users that keep producer prices below marginal production costs. These policies are not shown in the graphical profiles. Countries also apply production taxes, royalties and other levies on the extraction or harnessing of energy resources (such as crude oil royalties and water "rentals" paid by hydroelectricity generators). Since they are not directly imposed on the energy product, and the rate to which they are passed through to end-users can vary largely, these taxes (and related tax expenditures) are not included in the graphical profiles. Furthermore, in some cases royalties and production taxes apply to internationally traded goods, which limits their impact on prices in the domestic market. Similarly, export taxes, which do not directly affect domestic prices of energy, may indirectly change domestic energy prices. These taxes (and related tax expenditures) are equally not included in the graphical profile. In order to provide some insight into the variety of policy measures that affect producer prices of energy, a summary of pricing measures applied in the countries analysed is presented alongside their graphical profiles and the main policies affecting producer prices are discussed in Part 1, Section 1.4.1 of this report.

Taxes aimed at emissions that are not fixed on a per-volume-of-fuel basis are not included, since the emissions vary according to the quality of the fuel and the combustion technology used. This includes, for example, taxes on directly measured emissions from fuel combustion such as those on $NO_{X,}$ or SO_X emissions, and taxes based on the amount of sulphur (which produces SO_X) per fuel volume.

Tax rates, which are set in monetary units per physical quantity of fuel, or as *ad valorem* taxes, translated into per-unit rates, are re-calculated as effective tax rates per GigaJoule of energy (in the first map for each country) and per tonne of CO_2 emissions (in the second map). Tax rates are shown in local currency on the left-hand axis of the maps, and in euros (converted by reference to the average exchange rate over the 12 months ending August 2012). An alternative approach would convert the rates using purchasing power parity exchanges rates, but ideally these would be based on a comparison of prices of a basket of energy products, since relative price levels for energy products may differ significantly from general price levels. The tax rate applying to each fuel is mapped on the graph as a shaded bar across the portion of energy use or carbon emissions (the tax base) to which the particular rate applies. The shaded area is thus an approximation of the revenue raised by the tax – the rate multiplied by the base.

In some cases, due to scale issues, approximations were needed to allow the tax system to be shown on the graph. In particular, some minor variations in tax schedules were not included (such as exemptions for fuel for emergency vehicles) and tax bases that were very small, have sometimes been included in other categories by taking a weighted average of the applicable tax rates (for example, fuel oil and diesel oil have sometimes been included within a single category). More information on such approximations can be found in the country-specific chapters.

For federal countries, given that the horizontal axis shows national energy use, the horizontal bars showing the tax rate applicable to a given range of energy use reflect only federal taxes. It would not be practical to accurately illustrate in a national graph taxation at the sub-national level (e.g., by provinces and states). This is because each category along the horizontal access would in principle need to have a subdivision for each sub-national jurisdiction, or at least each jurisdiction that has a different tax rate. Nonetheless, in order to give some sense of the range in the levels of taxes levied by sub-national jurisdictions, for federal countries where this is relevant, the graph indicates the combined level of federal and sub national tax in selected sub-national jurisdictions.

Treatment of electricity

Taxes levied on electricity consumption have been mapped as effective taxes on the fuels used to generate the electricity. In cases where a common nominal tax rate is applied to all electricity consumption, the effective tax rate on each underlying energy source (e.g., coal, natural gas, hydro) used to generate the electricity is shown. In cases where different rates of nominal electricity tax apply to consumption in different sectors (e.g., residential, commercial, industrial), for each sector, the effective tax rate shown is that on the "average" basket of fuels used to generate electricity in the country. Effective carbon tax rates on electricity need to be interpreted carefully because of this assumption. If carbon energy is a small proportion of the generation mix, the effective tax rate on carbon thus calculated will be very high.

As described with reference to tax bases, in cases where tax rates on electricity consumption are shown in the maps, they have been adjusted downward to take into account that the effective tax rate on the energy value of fuels used in producing electricity is lower than the nominal rate on the energy value of electricity consumed because of energy losses in generation and transmission.

Data sources

Tax rates are the rates in effect on 1 April 2012 unless otherwise indicated. Tax rate information (used to construct the vertical axes of the maps) was obtained from several sources. Key sources that were used for multiple countries include the *Database on Instruments Used for Environmental Policy and Natural Resources Management* (OECD/EEA), the *Inventory of Estimated Budgetary Support and Tax Expenditures for Fossil Fuels* (OECD, 2014), Kojima (2013) and in some cases from country analysis briefs by the Energy Information Administration (EIA, several years). Country-specific sources are detailed in the country-specific chapters. Bilateral consultations were also an important source in obtaining or confirming tax rates.

Tax expenditures

In addition to showing taxes levied, the maps developed in this paper can also illustrate tax exemptions, refunds and credits, reduced rates and other tax expenditures that reduce the effective tax rates for some portion of the tax base. A tax expenditure is a measure that reduces the tax burden relative to some benchmark level that would otherwise apply under the "normal" rules of the particular country.

The graphical profiles included in the country sections show tax expenditures for the products and users where these are reported by the countries concerned. They also show tax rebates and tax credits which apply to energy products, whether or not these are identified as tax expenditures *per se*. This is important in order for the graphical profiles to show the net level of tax on each category of energy use. Where more than one benchmark is encompassed within a single category in the maps (e.g., if the applicable tax bases are very small) a weighted average of the relevant benchmarks has been used.

Many countries estimate tax expenditures using the so-called revenue foregone method, which measures the tax expenditure as the rate of the tax concession multiplied by the relevant base or uptake. Such estimates do not take into account changes in behaviour that might occur if the measure were to be removed, which could affect the base of the tax expenditure and therefore the amount of revenue that would be raised by its removal. This is a specific implication of the fact that tax rates (the vertical axis) and tax bases (the horizontal axis) are not independent – changes in tax rates affect behaviour and therefore the size of tax bases.

In the energy use and taxation maps, the shaded area under the tax rate for any portion of the tax base is therefore an approximation of the revenue obtained from that particular tax base, or in the case of tax expenditures, the revenue foregone. These values, however, will generally differ somewhat from official estimates of tax revenues and tax expenditures – including the values in the *Inventory of Estimated Budgetary Support and Tax Expenditures for Fossil Fuels* (OECD, 2014) – for a number of reasons:

- tax rates and tax expenditures in the maps are generally based on parameters as of 1 October 2011 or 1 April 2012, while energy use data is for calendar year 2009;
- inexact matching between actual tax bases and energy use data (in terms of fuels, uses or users) and the omission from the maps of measures with very small tax bases;
- tax rates which are set in *ad valorem* terms and have been subsequently converted to per-unit rates using price information;
- uncertainty in some cases about benchmarks used for tax expenditure estimates.

Data sources

The tax expenditures reported by countries and related benchmarks were obtained from the *Inventory of Estimated Budgetary Support and Tax Expenditures for Fossil Fuels* (OECD, 2014), from national official budget data and government reports. Tax expenditures for Argentina were also retrieved from the tax expenditure report from 2012 to 2014 by the Argentinean Ministry of Finance (2012). Other country-specific sources are detailed in the relevant country section. In many cases, the benchmark applied by governments in calculating the impact of various different exemptions or refunds, which is not always explicit in public documents, has been confirmed in bilateral consultations.

Conversion of tax bases and tax rates to energy content and CO_2 emissions

To show tax base and rate information on a single graph, both the tax rates and tax bases presented in the maps were first converted from their original units (tax rates, for example, are often based on volume or weight) to energy content measured in GigaJoules and then to tonnes of CO_2 emissions.

Because of the differing energy and emission characteristics of different fuels, if the same tax rate is applied to two different fuels on a, for example, per litre basis, they would not generally have the same effective tax rate in energy or carbon content terms. If the policy aim is to tax two fuels consistently based on their energy or carbon content, a different nominal tax rate will be needed for the two fuels if the rate is set on a volume or mass basis.

In the electricity sector, the carbon content refers to the carbon content of the underlying primary energy sources used to generate the electricity.

Coverage of CO_2 emissions

Among greenhouse gases, account is only taken of CO_2 emissions, which constitute about 82% of total greenhouse gases emissions in OECD countries (IEA, 2009). Other greenhouse gas emissions have not been included for two reasons:

- The level of CO_2 emissions from fuel combustion is fixed for a given quantity of each carbon fuel regardless of the combustion technology used. By contrast, the emissions factors for other greenhouse gases like methane (CH_4) and nitrous oxide (N_2O) change significantly depending on the technology used in the combustion process (IPCC, 2006). These two pollutants account for 15% of GHG emissions in the OECD area. The three remaining GHG gases [hydrofluorocarbons (HFC), perfluorocarbons (PFC) and sulphur hexafluoride (SF6)] are produced by specific industrial processes (e.g., primary aluminium production for PFC) rather than by fuel combustion.
- Among the countries examined, where there is a component of energy taxes that reflects greenhouse gases, it is in fact related exclusively to CO_2 rather than to any of the greenhouse gases associated with energy combustion.

The CO_2 emissions shown in this paper may differ from those provided in other data sources (such as country GHG inventories for the UNFCC for three main reasons:

- Firstly, consistent with the focus of this report on energy, only emissions relating to fuel combustion are shown; those from other sources, such as chemical reactions in industrial processes, landfills, agriculture land and livestock are not shown.
- Secondly, biofuels are treated as having a positive level of carbon emissions in these maps, as discussed below.
- Finally, the carbon emissions shown on the graphs are calculated using information on energy consumption for all the various fuels for each country from the International Energy Agency (IEA), converted to carbon emissions using standard conversion factors from the Intergovernmental Panel on Climate Change and the IEA as set out above.

The emissions figures, therefore, are essentially a "bottom-up" estimation of the carbon emissions from fuel use.

Emissions from biomass

There are three ways that CO_2 emissions from combustion of biomass (plants and trees) could potentially have been treated for the purposes of the maps:

- As with fossil fuels (coal, oil, natural gas), include direct human-made CO_2 emissions from biomass combustion, which would lead to positive emissions factors.

- Take into account the fact that plants and trees absorb CO_2 from the atmosphere as they grow, and emit CO_2 back into the atmosphere when they die – whether by human-made combustion or natural causes such as forest fires or decomposition – and relatively short life-cycle of plants and trees (years, decades or centuries) as compared with fossil fuels (millions of years). On this basis, the net effect of biomass production and consumption is sometimes considered to be "carbon neutral" as a general principle, leading to a zero emission factor.
- Estimate the emissions implied by particular types of biomass by following a life-cycle approach that examines the source of the biomass ("natural" vs. human-initiated crops), impacts of land use change, whether and how harvested biomass is replanted, etc. Under this approach, the carbon balance of biomass is neutral only if the original biomass is replanted in the same form after harvest, thus preserving its carbon absorption (sink) function. Depending on the particular source, this approach could potentially lead to positive or negative emission factors.

Considering biomass as automatically carbon neutral (the second option) is likely to be too simplistic. Assuming zero net emissions involves making assumptions about the lifecycle of biomass that may not be accurate. On the other hand, adopting a life-cycle approach (third option) to obtain an overall estimate of the carbon impact of biomass production from particular sources in particular countries is beyond the scope of this report and would be very challenging to do even as a stand-alone project, given the heterogeneity of biomass sources and the limited information available.

For these reasons, in order to provide the most information, the first option was followed, so that the direct emissions produced by biomass combustion are shown, in the same way as with fossil energy. Users who are comfortable that a particular biomass source can be considered carbon neutral can choose to ignore those carbon values.

Data sources

Information from the Extended World Energy Balances of OECD countries (IEA, 2014) and Oil Information (IEA, 2011) was used to convert volume- or mass-based tax rates or consumption data to GigaJoules.

The emission factors were taken from the Intergovernmental Panel on Climate Change's *Guidelines for National Greenhouse Gas Inventories* (IPCC, 2006) and applied to the corresponding energy products. The information used is consistent with the emission factors listed by the IEA in its report *CO_2 Emissions from Fuel Combustion* (IEA, 2011). In some cases, a greater level of detail was available in the IPCC publication, and for this reason it has been used as the primary data source. The IPCC provides the data which allow the conversion from GigaJoules to carbon content. Carbon content data have been converted to CO_2 by using the molecular weight ratio of CO_2 to C (44/12).

Note

1. This gross-up is based on the observed energy efficiency of electricity generation in the country in question as derived from the IEA energy balances. In the case of combined heat and power (also known as cogeneration) plants, which produce both electricity and heat using the same fuel, it is assumed that heat is a by-product and is 100% efficient. The energy value of the marketable heat output is subtracted from the energy value of fuel inputs. The remaining fuel use is attributed to electricity generation, from which an effective efficiency rate can be calculated.

ANNEX B

Effective tax rate – Overview by country

		EUR per GJ						EUR per tonne CO_2				
		Oil products	Coal and peat	Natural gas	Combustibles and waste	Renewables	Total	Oil products	Coal and peat	Natural gas	Combustibles and waste	Total
ARG	*Road*	10.84	0.00	1.71	5.04	0.00	9.18	150.54	0.00	30.55	71.18	132.79
	Non-road	4.06	0.00	0.02	0.00	0.00	2.49	56.44	0.00	0.33	0.00	37.78
	Total transport	10.45	0.00	1.46	5.04	0.00	8.68	145.14	0.00	25.99	71.18	125.95
	Residential and *commercial*	1.15	0.00	0.24	0.00	0.00	0.37	17.55	0.00	4.24	0.00	6.31
	Industrial and *energy transformation*	0.15	0.00	0.11	0.00	0.00	0.12	2.09	0.00	1.93	0.00	1.74
	Total heating and process	0.30	0.00	0.16	0.00	0.00	0.20	4.20	0.00	2.84	0.00	3.00
	Electricity	0.10	0.09	0.31	0.07	0.19	0.24	1.30	0.82	5.47	0.64	3.81
	Total	4.71	0.05	0.30	0.13	0.19	1.84	64.76	0.38	5.43	1.20	29.92
BRA	*Road*	0.00	0.00	0.00	0.00	0.00	0.00	0.00	0.00	0.00	0.00	0.00
	Non-road	0.00	0.00	0.00	0.00	0.00	0.00	0.00	0.00	0.00	0.00	0.00
	Total transport	0.00	0.00	0.00	0.00	0.00	0.00	0.00	0.00	0.00	0.00	0.00
	Residential and *commercial*	0.00	0.00	0.00	0.00	0.00	0.00	0.00	0.00	0.00	0.00	0.00
	Industrial and *energy transformation*	0.00	0.00	0.00	0.00	0.00	0.00	0.00	0.00	0.00	0.00	0.00
	Total heating and process	0.00	0.00	0.00	0.00	0.00	0.00	0.00	0.00	0.00	0.00	0.00
	Electricity	0.76	0.62	0.89	0.89	1.81	1.56	10.24	4.68	15.79	7.90	8.35
	Total	0.03	0.15	0.13	0.05	1.80	0.35	0.41	1.26	2.40	0.46	0.62
CHN	*Road*	3.23	0.00	0.00	0.00	0.00	3.19	45.17	0.00	0.00	0.00	44.64
	Non-road	1.46	0.00	0.00	0.00	0.00	1.33	19.94	0.00	0.00	0.00	17.64
	Total transport	2.87	0.00	0.00	0.00	0.00	2.78	39.80	0.00	0.00	0.00	38.38
	Residential and *commercial*	1.49	0.00	0.00	0.00	0.00	0.20	21.40	0.00	0.00	0.00	2.09
	Industrial and *energy transformation*	1.43	0.00	0.00	0.00	0.00	0.18	20.00	0.00	0.00	0.00	1.79
	Total heating and process	1.45	0.00	0.00	0.00	0.00	0.19	20.43	0.00	0.00	0.00	1.88
	Electricity	2.18	0.00	0.00	0.00	0.00	0.01	29.69	0.00	0.00	0.00	0.13
	Total	2.18	0.00	0.00	0.00	0.00	0.31	30.48	0.00	0.00	0.00	3.40
IND	*Road*	2.35	0.00	0.00	0.00	0.00	2.24	32.38	0.00	0.00	0.00	31.15
	Non-road	1.09	0.00	0.00	0.00	0.00	1.09	14.76	0.00	0.00	0.00	14.76
	Total transport	2.21	0.00	0.00	0.00	0.00	2.12	30.45	0.00	0.00	0.00	29.41
	Residential and *commercial*	0.00	0.04	0.00	0.00	0.00	0.00	0.00	0.43	0.00	0.00	0.02
	Industrial and *energy transformation*	1.06	0.04	0.00	0.00	0.00	0.29	15.40	0.35	0.00	0.00	3.15
	Total heating and process	0.68	0.04	0.00	0.00	0.00	0.14	9.98	0.36	0.00	0.00	1.43
	Electricity	1.55	0.04	0.00	0.00	0.00	0.10	20.86	0.44	0.00	0.00	1.19
	Total	1.36	0.04	0.00	0.00	0.00	0.28	19.33	0.41	0.00	0.00	3.12
IDN	*Road*	0.00	0.00	0.00	0.00	0.00	0.00	0.01	0.00	0.00	0.00	0.01
	Non-road	0.00	0.00	0.00	0.00	0.00	0.00	0.00	0.00	0.00	0.00	0.00
	Total transport	0.00	0.00	0.00	0.00	0.00	0.00	0.01	0.00	0.00	0.00	0.01
	Residential and *commercial*	0.00	0.00	0.00	0.00	0.00	0.00	0.00	0.00	0.00	0.00	0.00
	Industrial and *energy transformation*	0.00	0.00	0.00	0.00	0.00	0.00	0.00	0.00	0.00	0.00	0.00
	Total heating and process	0.00	0.00	0.00	0.00	0.00	0.00	0.00	0.00	0.00	0.00	0.00
	Electricity	0.00	0.00	0.00	0.00	0.00	0.00	0.00	0.00	0.00	0.00	0.00
	Total	0.00	0.00	0.00	0.00	0.00	0.00	0.01	0.00	0.00	0.00	0.00
RUS	*Road*	0.00	0.00	0.00	0.00	0.00	0.00	0.05	0.00	0.00	0.00	0.05
	Non-road	0.00	0.00	0.00	0.00	0.00	0.00	0.01	0.00	0.00	0.00	0.00
	Total transport	0.00	0.00	0.00	0.00	0.00	0.00	0.04	0.00	0.00	0.00	0.03
	Residential and *commercial*	0.00	0.00	0.00	0.00	0.00	0.00	0.02	0.00	0.00	0.00	0.00
	Industrial and *energy transformation*	0.00	0.00	0.00	0.00	0.00	0.00	0.02	0.00	0.00	0.00	0.00
	Total heating and process	0.00	0.00	0.00	0.00	0.00	0.00	0.02	0.00	0.00	0.00	0.00
	Electricity	0.00	0.00	0.00	0.00	0.00	0.00	0.00	0.00	0.00	0.00	0.00
	Total	0.00	0.00	0.00	0.00	0.00	0.00	0.03	0.00	0.00	0.00	0.01

		EUR per GJ						EUR per tonne CO_2				
		Oil products	Coal and peat	Natural gas	Combustibles and waste	Renewables	Total	Oil products	Coal and peat	Natural gas	Combustibles and waste	Total
ZAF	*Road*	8.75	0.00	0.00	0.00	0.00	8.75	122.65	0.00	0.00	0.00	122.65
	Non-road	0.54	0.00	0.00	0.00	0.00	0.54	7.59	0.00	0.00	0.00	7.59
	Total transport	8.17	0.00	0.00	0.00	0.00	8.17	114.56	0.00	0.00	0.00	114.56
	Residential and *commercial*	0.43	0.00	0.00	0.00	0.00	0.04	5.98	0.00	0.00	0.00	0.35
	Industrial and *energy transformation*	3.10	0.00	0.00	0.00	0.00	0.24	45.51	0.00	0.00	0.00	2.60
	Total heating and process	2.28	0.00	0.00	0.00	0.00	0.18	32.87	0.00	0.00	0.00	1.90
	Electricity	8.79	0.33	0.00	0.00	0.30	0.33	118.58	3.45	0.00	0.00	3.46
	Total	6.90	0.19	0.00	0.00	0.29	1.25	97.20	2.02	0.00	0.00	13.86

StatLink http://dx.doi.org/10.1787/888933206022

ORGANISATION FOR ECONOMIC CO-OPERATION AND DEVELOPMENT

The OECD is a unique forum where governments work together to address the economic, social and environmental challenges of globalisation. The OECD is also at the forefront of efforts to understand and to help governments respond to new developments and concerns, such as corporate governance, the information economy and the challenges of an ageing population. The Organisation provides a setting where governments can compare policy experiences, seek answers to common problems, identify good practice and work to co-ordinate domestic and international policies.

The OECD member countries are: Australia, Austria, Belgium, Canada, Chile, the Czech Republic, Denmark, Estonia, Finland, France, Germany, Greece, Hungary, Iceland, Ireland, Israel, Italy, Japan, Korea, Luxembourg, Mexico, the Netherlands, New Zealand, Norway, Poland, Portugal, the Slovak Republic, Slovenia, Spain, Sweden, Switzerland, Turkey, the United Kingdom and the United States. The European Union takes part in the work of the OECD.

OECD Publishing disseminates widely the results of the Organisation's statistics gathering and research on economic, social and environmental issues, as well as the conventions, guidelines and standards agreed by its members.

OECD PUBLISHING, 2, rue André-Pascal, 75775 PARIS CEDEX 16
(23 2015 12 1P) ISBN 978-92-64-23232-7 – 2015

www.ingramcontent.com/pod-product-compliance
Lightning Source LLC
LaVergne TN
LVHW081404110826
845149LV00010B/1658
9789264232327